# R Programming
# – An Approach to Data Analytics

**MJP**
PUBLISHERS

# R Programming
# – An Approach to Data Analytics

**Dr. G. Sudhamathy**

Assistant Professor, Department of Computer Science,
Avinashilingam University, Coimbatore – 641043.

**Dr. C. Jothi Venkateswaran**

Principal, Government Arts and Science College,
Perumbakkam, Chennai – 600100.

Chennai          New Delhi          Tirunelveli

**MJP PUBLISHERS**

ISBN 978-81-8094-408-6    **MJP Publishers**

All rights reserved    No. 44, Nallathambi Street,
Printed and bound in India    Triplicane, Chennai 600 005

MJP   371    © Publishers, 2018

Publisher:   C. Janarthanan

Project Editor:   C. Ambica

# FOREWORD

If you are looking for a complete step-by-step instructions for learning R Programming for Statistical Data Analysis, Graphical Visualization and Data Mining, authors Dr. Sudhamathy & Dr. Jothi venkateswaran's "R Programming - An Approach to Data Analytics" is a hands-on book packed with examples and references that would help you get started coding in R for variety of data science problems.

As the authors explain in their book, understanding the techniques and algorithms of data analytics for large dataset is critical for effective data classification. This helps as developer not just learn R Programming but also to apply right algorithms and statistical model.

Hopefully you can take the instructions provided in this book to get started in R programming for your next data analysis project, do some exciting data visualization and data mining on your own.

**Mr. Sajeev Madhavan**

*Director of Architecture, **ORACLE**,*

*San Francisco Bay Area, USA*

# FOREWORD

It's my immense happiness in penning this foreword for a book that is quite impressive for any techie who is interested in R-programming. It's also equally joyous to have a book written by experts, Dr. G. Sudhamathy and Dr. C. Jothi Venkateswaran. When a book can teach you and guide you as you work hands on the tool, you are in the right direction in your learning path.

This book on R-Programming starts with simple concepts like programming logic statements, data types, moves on to cover some advanced topics like Statistical Analysis using R, Data mining using R, in addition to the Graphics programming. The concepts are well supported by the real-time works carried out in R, with sufficient figurative illustrations wherever it is strongly essential for explanations. The powerful presentations on the Analytics chapters have an in-depth knowledge of the tool, with a vital emphasis on the Analytics which have been supported with the five most important case-studies given towards the end of the book.

One can be definitively sure that book will be of great help and guidance for the learner to carry out their works on Analytics using R, either in the research, practice or just to learn the tool.

My heartfelt appreciations to the authors who have done a wonderful job in bringing out this book which is very much needed at this point of the hour where the entire world is diving into data, Big data and analyzing those data for newer knowledge and perceptions to drive everyday business.

Best wishes for a bestselling of this book in the Academia, Research and Practice.

**Dr. S. Justus**

*Associate Professor & Chair – Software Engineering Research Group*
*VIT University, Chennai*

# PREFACE

Huge volumes of data are being generated by many sources like commercial enterprises, scientific domains and general public daily. According to a recent research, data production will be 44 times greater in 2020 than it was in 2010. Data being a vital resource for business organizations and other domains like education, health, manufacturing etc., its management and analysis is becoming increasingly important. This data, due to its volume, variety and velocity, often referred to as Big Data, also includes highly unstructured data in the form of textual documents, web pages, graphical information and social media comments. Since Big Data is characterised by massive sample sizes, high dimensionality and intrinsic heterogeneity, traditional approaches to data management, visualisation and analytics are no longer satisfactorily applicable. There is therefore an urgent need for newer tools, better frameworks and workable methodologies for such data to be appropriately categorised, logically segmented, efficiently analysed and securely managed. This requirement has resulted in an emerging new discipline of Data Science that is now gaining much attention with researchers and practitioners in the field of Data Analytics.

R is a programming language and a free open source software environment for data analytics. It is growing exponentially by most measures, count over a million users, and it has over 10,865 standard and add-on packages contributed by the community, with that number increasing by about 25% each year. R is a powerful tool for approaching statistical, graphical, and data mining problems. It is used by many organizations and individuals daily to perform serious data analytics. R does not require the users to have basic programming knowledge as it is made up of many inbuilt packages and function which can achieve very complex processing easily in fraction of seconds. This book is full of easy simple steps to achieve greater results with complex data. It also details on how to model a specific problem and come out with predictions for the future. The main motivation of this book is to break the complexities remaining in the minds of students and researchers about

R programming language and make it easy to approach by any one. The chapters are designed in such a fashion that it targets the beginners with the first 4 chapters and targets the advanced concept learners in the next 3 chapters. The book also helps the reader with the list of all packages and functions used in this book along with the page numbers to know the usage of those. Every concept discussed in the various sections in this book has proper example dealt with a set of code and its results (as text or as graphs).

The book is organized into 7 chapters and the concept discussed in each chapter is as detailed below.

**Chapter 1** introduces the programming language R, briefs on how to install R Studio, how to use the editor and write simple code using R. This chapter also details on how to get help in R from its manuals and how to perform simple mathematical operations using R. The chapter then progresses with the introduction of the concepts of packages, environments and functions. Finally this chapter details on the programming concepts of flow control and loops.

**Chapter 2** discusses on the basic data types in R, the primitive data types such as vectors, matrices and arrays, lists and factors. It also deals with the complex data types such as data frames, strings, dates and times. The chapter not only discusses on the data creation, but also basic operations on the data of different data types.

**Chapter 3** deals with data preparation in which it details on how and where to fetch the datasets from, how to import and export data from various sources which are of different types like CSV files, XML files, etc. It also discusses on the ways of accessing various databases. The data cleaning and transformation techniques such as data reshaping, grouping functions are also outlined in this chapter.

**Chapter 4** is about using the graphical features in R for exploratory data analysis. It gives examples of pie charts, scatter plots, line plots, histograms, box plots and bar plots using the various graphical packages such as base, lattice and ggplot2.

**Chapter 5** deals with statistical analysis concepts using R such as the basic statistical measures like mean, median, mode, standard deviation, variance and ranges. It discusses on the distribution of data as normal distribution and binomial distribution and how it can be viewed and analyzed using R. Then, the chapter explores on the complex statistical techniques such as correlation analysis, regression analysis, ANOVA and hypothesis testing which can be implemented using R.

**Chapter 6** is about exploring the data mining techniques available in R. It explores the K-Means, K-Medoids, Hierarchical and Density Based Clustering techniques using proper examples and case studies. The decision tree classification techniques are also discussed with suitable examples. Outlier detection is also explored using various techniques such as univariate, multivariate, LOF and clustering. Dimensionality reduction is done using PCA and feature selection. Association rule mining is done using the titanic dataset and a proper case study analysis is presented in this chapter.

**Chapter 7** is mainly to explore the various essential case studies such as text analytics, credit risk analysis, social network analysis and few exploratory data analysis. The main purpose of this chapter is to use the basic and advanced concepts presented in the other previous chapters of this book.

# ACKNOWLEDGEMENTS

One of the authors (Dr. G. Sudhamathy) thanks the authorities of Avinashilingam University, Coimbatore, for providing all the support for making this book a reality.

The author expresses her reverential gratitude to Shri. Dr. P. R. Krishnakumar, Chancellor, Dr. Premavathy Vijayan, Vice Chancellor and Dr. A. Kowsalya, Registrar, Avinashilingam Universty, Coimbatore, for providing the opportunity to publish this book.

The author would like to mention her special regards and thanks to Dr. G. P. Jeyanthi, Research and Consultancy Director, Dr. A. Parvathi, Dean, Faculty of Science and Dr. V. Radha, Head, Department of Computer Science, Avinashilingam Universty, Coimbatore, for their constant encouragement and support to turn this work into a useful product.

A special thanks to Dr. G. Padmavathi, Professor, Department of Computer Science, Avinashilingam University, Coimbatore, who was the motivational support for acquiring the technical knowledge behind this book.

The author wishes to thank all the faculty members of the Department of Computer Science, Avinashilingam University, Coimbatore, for their continuous support and suggestions for this book.

We are grateful to the students and teacher community who kept us on our toes with their constant bombardment of queries which prompted us to learn more, simplify our learning and findings and place them neatly in a book.

Our Special regards for the experts Mr. Sajeev Madhavan, Director of Architecture, Oracle, USA and Dr. S. Justus, Associate Professor, VIT, Chennai who gave their expert opinion in shaping this book into a more appealing format.

Most importantly we would like to thank our family members without whose support this book would not have been a reality.

Last, but not the least, this work is a dedication to God, the Almighty whose grace has showered upon us in making our dream come true.

**G. Sudhamathy**

**C. Jothi Venkateswaran**

# CONTENTS

# CHAPTER 1

# BASICS OF R

❖ **OBJECTIVES**

On completion of this Chapter you will be able to:

- understand how R is different from other languages
- install R in your system
- write a beginners program using R
- get help in R
- assign variables in R
- know the basic mathematical operations in R
- understand various environments and scope of variables
- understand functions in R
- understand program flow control in R
- understand loops in R

## 1.1. Introducing R

R is a Programming Language and R also refers to the software that is used to run the R programs. Ross Ihaka and Robert Gentleman from University of Auckland created R language in 1990s. R language is based on the S language. S Language was developed at the Bell Laboratories in 1970s. S Language was developed by John Chambers. R Software is a GNU project free and open source software. R (Language and Software) is developed by the R Core Team. R has evolved over the past 3 to 4 decades as its history originated from 1970s.

One can write a new package in R if the existing package is not sufficient for the individual's use. R is a high-level scripting language which need not be compiled, but it is an interpreted language. R is an imperative language and still it supports object-oriented programming.

R is a free open source language that has cross platform compatibility. R is a most advanced statistical programming language and it can produce outstanding graphical outputs. R is extremely flexible and comprehensive even for the beginners. R easily relates to other programming languages such as C, C++, Java, Python, Hadoop, etc. R can handle huge data in flat files even in semi structured or in unstructured form.

The R language allows the user to program loops to successively analyze several data sets. It is also possible to combine in single program different statistical functions to perform more complex analyses. The R users may get benefitted from a large number of programs written and available on the internet. At first R can look very complex for a beginner or non-specialist. But, this is not actually true as the prominent feature of R is its flexibility. R displays the results of the analysis immediately and these results are stored in "objects" so that further analysis can be done on them. The user can also extract a part of the result which is of interest to him.

Looking at the features of R, some users may think that "I can't write programs using R". But, this is not the case for two reasons. First, R is an interpreted language and not a compiled one. This means that all commands typed on the keyboard are directly executed without need to build the complete program like in C, C++ or Java. Second, R's syntax is very simple and intuitive.

In R, a function is always written with parentheses, eg. *ls()*. If only the name of the function is typed, R displays the content of the function. In this book the functions are written with their names followed by parentheses to distinguish them from other objects. When R is running variables, data, functions, results, etc. are stored in the active memory of the computer in the form of *objects* which have a *name*. The user can do actions on these *objects* with *operators* and *functions*.

## 1.2. Installing R

R is available in several forms, essentially for Unix and Linux machines, or some pre-compiled binaries for Windows, Linux and Macintosh. The files needed to install R, either from the source or from the pre-compiled binaries are distributed from the internet site of the Comprehensive R Archive Network (CRAN) where the instructions for installation are also available.

R can be installed from the link http://www.r-project.org using internet connection. Use the "Download R" link in web page to download the R Executable. Choose the version of R that is suitable for your operating system. R-Scripts can run without the installation of the IDE, the R-Studio using the R-Console. The prerequisite for installing R-Studio is that one should have downloaded and installed any version of R. (Version 3.3.0 of R is used for installation and running scripts used in this book). Follow the instructions on the website to complete installation of R Console.

Once R installation is completed we install R-Studio. For installation of R-Studio in Windows operating system, we download the latest precompiled binary distribution from the CRAN website http://www.rstudio.org. (Version 3.4 of R-Studio is used for installation and running scripts used in this book). Start the installation and follow the steps required by the setup wizard. Once completed, launch RStudio IDE from Start à All Programs à Rstudio à RStudio.exe or from your custom installation directory. The default installation directory for RStudio IDE is "C:\Program Files\RStudio\bin\rstudio.exe.

R Studio is an Integrated Development Environment (IDE) that consists of a GUI with four parts – 1) A text editor 2) command-line interpreter 3) place to display files, plots, packages and help information 4) place to display the data being used and the variables used in the program (Environment / History).

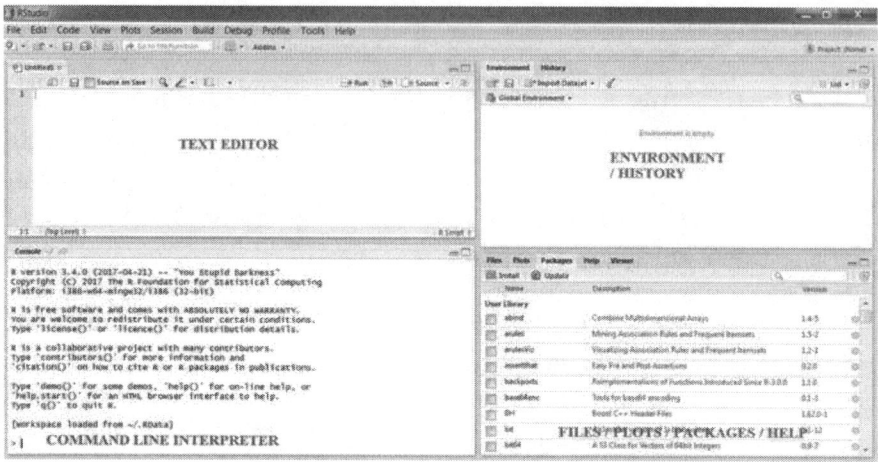

**Figure 1.1**   R-Studio GUI

# 1.3. Initiating R

## 1.3.1. First Program

Open R Gui, find the command prompt and type the command below and hit enter to run the command.

> *sum(1:5)*

*[1] 15*

The result above shows that the command gives the result 15. That is the command has taken the input of integers from 1 to 5 and has performed the sum operation on them. In the above command *sum()* is a *function* that takes the argument *1:5* which means a *vector* that consists of a sequence of integers from 1 to 5. Like any other command prompt, R also allows to use the up arrow key to revoke the previous commands.

## 1.3.2. Help in R

There are many ways to get help from R. If a function name or a dataset name is known then we can type ? followed by the name. If name is not known then we need to type **??** followed by a term that is related to the search function. Keywords, special characters and two separate terms of search need to be enclosed in double or single quotes. The symbol **#** is used to comment a line in R Program like any other programming language.

```
> ?mean                   # help page for mean function opens
> ?"+"                     # help page for addition function opens
> ?"if"                    # help page for if opens
> ??plotting               # searches for the help pages containing the word "plotting"
> ??"regression model"     # searches for "regression model" phrase
```

The same help can be obtained by the functions *help()* and *help.search()*. In these functions the arguments has to be enclosed by quotes.

```
> help("mean")
> help("+")
> help("if")
> help.search("plotting")
> help.search("regression model")
```

## 1.3.3. Assigning Variables

The results of the operations in R can be stored for reuse. The values can be assigned to the variables using the symbol "<-" or "=" of which the symbol "<-" is preferred. There is no concept of variables declaration in R. The variable type is assumed based on the value assigned.

```
> X <- 1:3
> X
[1] 1 2 3
> Y = 4:6
> Y
[1] 4 5 6
> X + 3 * Y - 2
[1] 11 15 19
```

The variable names consist of letters, numbers, dots and underscores, but a variable name should only start with an alphabet. The variable names should not be reserve words. To create global variables (variables available everywhere) we use the symbol "<<-".

$X <<- exp(exp(1))$

Assignment operation can also be done using the *assign()* function. For global assignment the same function *assign()* can be used, but, by including an extra attribute *globalenv()*. To see the value of the variable, simply type the variable in the command prompt. The same thing can be done using a *print()* function.

> *assign("F", 3 * 8)*

> *assign("G", 6 * 9, globalenv())*

> *F*

*[1] 24*

> *print(G)*

*[1] 54*

If assignment and printing of a value has to be done in one line we can do the same in two ways. First method, by separating the two statements by a semicolon and the second method is by wrapping the assignment in parenthesis () as below.

> *L <- sum(4:8); L*

*[1] 30*

> *(M <- sum(5:9))*

*[1] 35*

## 1.3.4. Basic Mathematical Operations

The "+" plus operator is used to perform the addition operation. It can be used to add two numbers or add two vectors. Vector represents an ordered set of values. Vectors are mainly used to analyse statistical data. The ":" colon operator creates a sequence. Sequence is a series of numbers within the given limits. The "c()" function concatenates the values given within the brackets "(" and ")". Variable names in R are case sensitive. Open R Gui, find the command prompt and type the command below and hit enter to run the command.

> > 7:12 + 12:17
>
> [1] 19 21 23 25 27 29
>
> > c(3, 1, 8, 6, 7) + c(9, 2, 5, 7, 1)
>
> [1] 12  3 13 13  8

The vectors and the *c()* function in R helps us to avoid loops. The statistical functions in R can take the vectors as input and produce results. The *sum()* function takes vector arguments and produces results. But, the *median()* function when taking the vector arguments shows errors.

> > sum(7:10)
>
> [1] 34
>
> > mean(7:10)
>
> [1] 8.5
>
> > median(7:10)
>
> [1] 8.5
>
> > sum(7,8,9,10)
>
> [1] 34
>
> > mean(7,8,9,10)
>
> [1] 7
>
> > median(7,8,9,10)
>
> *Error in median(7, 8, 9, 10) : unused arguments (9, 10)*

Similar to the "+" plus operator all other operators in R take vectors as inputs and can produce results. The subtraction and the multiplication operations work as below.

> > c(5, 6, 1, 9) - 2
>
> [1]  3  4 -1  7
>
> > c(5, 6, 1, 9) - c(4, 2, 0, 7)

*[1] 1 4 1 2*

*> -1:4 \* -2:3*

*[1] 2 0 0 2 6 12*

*> -1:4 \* 3*

*[1] -3 0 3 6 9 12*

The exponentiation operator is represented using the symbol "^" or the "\*\*". This can be checked using the function *identical()*.

*> identical(2^3, 2\*\*3)*

*[1] TRUE*

The division operator is of three types. The ordinary division is represented using the "/" symbol, the integer division operator is represented using the "%/%" symbol and the modulo division operator is represented using the "%%" symbol. The below example commands show the results of the division operators.

*> 5:9/2*

*[1] 2.5 3.0 3.5 4.0 4.5*

*> 5:9%/%2*

*[1] 2 3 3 4 4*

*> 5:9%%2*

*[1] 1 0 1 0 1*

The other *mathematical functions* are the trigonometry functions like, *sin()*, *cos()*, *tan()*, *asin()*, *acos()*, *atan()* and the logarithmic and exponential functions like *log()*, *exp()*, *log1p()*, *expm1()*. All these mathematical functions can operate on vectors as well as individual elements. Few more examples of the mathematical functions are listed below

The operator "==" is used for comparing two values. For checking inequalities of values the operator "!=" is used. These operators are called the *relational operators*. The relational operators also take the vectors as input and operate on them. The other relational operators are the "< ", "> ", "<= " and ">= ".

*> c(2, 4 - 2, 1 + 1) == 2*

*[1] TRUE TRUE TRUE*

> 1:5 != 5:1

*[1] TRUE TRUE FALSE TRUE TRUE*

> exp(1:3) < 20

*[1] TRUE TRUE FALSE*

> (1:10) ^ 2 >= 50

*[1] FALSE FALSE FALSE FALSE FALSE FALSE FALSE TRUE TRUE TRUE*

Non-integers cannot be compared using the operator "==" as it produces wrong results due to rounding off error of the float numbers being compared. For overcoming this issue we have the function *all.equal()*. If the value to be compared by the function *all.equal()* is not equal, it returns a report on the difference. To get a TRUE or FALSE reply, the *all.equal()* function has to be wrapped using the function *isTRUE()*. The below example code will help to understand the concepts discussed.

> sqrt(2) ^ 2 == 2

*[1] FALSE*

> sqrt(2) ^ 2 - 2

*[1] 4.440892e-16*

> all.equal(sqrt(2) ^ 2, 2)

*[1] TRUE*

> all.equal(sqrt(2) ^ 2, 3)

*[1] "Mean relative difference: 0.5"*

> isTRUE(all.equal(sqrt(2) ^ 2, 3))

*[1] FALSE*

The equality operator "==" can also be used to compare strings, but, string comparison is case sensitive. Similarly, the operators "<" and ">" can also be used on strings. The below examples show the results.

> c("Week", "WEEK", "week", "weak") == "week"

*[1] FALSE FALSE TRUE FALSE*

```
> c("A", "B", "C") < "B"
[1]  TRUE FALSE FALSE
> c("a", "b", "c") < "B"
[1]  TRUE  TRUE FALSE
```

# 1.4. Packages in R

R Packages are installed in an online repository called CRAN (Comprehensive R Archive Network). A Package is a collection of R functions and datasets. Currently, the CRAN package repository features *10756 available packages*. The list of all available packages in the CRAN repository can be viewed from the web site *"https:// cran.r-project.org/web/packages/available_packages_by_name.html"*. To find the list of functions available in a package (say the package is *"stats"*) we can use the command *ls("package:stats")* or the command *library(help = stats)* in the command prompt.

A library is a folder in the machine that stores the files for a package. If a package is already installed on a machine we can load the same using the *library()* function. The name of the package to be loaded is passed to the *library()* function as argument without enclosing in quotes. If the package name has to be programmatically passed to the *library()* function, then we need to set the argument *character.only = TRUE*. If a package is not installed and if the *library()* function is used to load the package, it will throw an error message. Alternatively if the *require()* function is used to load a package, it returns TRUE if the package is already installed or it returns FALSE if the package is not already installed.

We can list and see all the packages that are already loaded using the *search()* function. This list shows the global environment as the first one followed by the recently loaded packages. The last two are special environments, namely, "Autoloads" and "base" package.

```
> search()
[1] ".GlobalEnv"      "package:cluster"  "tools:rstudio"    "package:stats"
[5] "package:graphics" "package:grDevices" "package:utils"    "package:datasets"
[9] "package:methods"  "Autoloads"        "package:base"
```

The function *installed.packages()* returns a data frame with information about all the packages installed in a machine. It is safe to view the results of this using the *View()* function as it may list hundreds of packages. This list of packages also shows the version of the package installed, location on the machine and dependent packages.

> *View(installed.packages())*

The function *R.home("library")* retrieves the location on the machine that stores all R default packages. The same result can be accomplished using the *.Library* command. The home directory can be listed using the *path.expand("~")* and *Sys. getenv("HOME")* functions.

> *R.home("library")*

[1] "C:/PROGRA~1/R/R-33~1.0/library"

> *.Library*

[1] "C:/PROGRA~1/R/R-33~1.0/library"

> *path.expand("~")*

[1] "C:/Users/admin/Documents"

> *Sys.getenv("HOME")*

[1] "C:/Users/admin/Documents"

When R is upgraded, it is required to reinstall all the packages as different versions of R needs different versions of the packages. The function *.libPaths()* lists all the R libraries in the installed machine. The first value listed is the place where the packages will be installed by default.

> *.libPaths()*

[1] "C:/Users/admin/Documents/R/win-library/3.3"

[2] "C:/Program Files/R/R-3.3.0/library"

The CRAN package repository contains handful of packages that needs special attention. To access additional repositories, type *setRepositories()* and select the repository required. The repositories R-Forge and rforge.net contains the development versions of the packages that appear on the CRAN repository. The function *available.packages()* lists thousands of packages in each of the selected repository. (*Note*: can use the *View()* function to restrict fetching of thousands of the packages at one go)

> *setRepositories()*

*--- Please select repositories for use in this session ---*

    *1: + CRAN*

    *2:  BioC software*

    *3:  BioC annotation*

    *4:  BioC experiment*

    *5:  BioC extra*

    *6:  CRAN (extras)*

    *7:  Omegahat*

    *8:  R-Forge*

    *9:  rforge.net*

    *10: + CRANextra*

*Enter one or more numbers separated by spaces, or an empty line to cancel*

*1:*

There are many online repositories like *GitHub, Bitbucket,* and *Google Code* from where many R Packages can be retrieved. The packages can be installed using the function *install.packages()* function by mentioning the name of the package as argument to this function. But, it is necessary to have internet connection to install any package and write permission to the hard drive. To update the latest version of the installed packages, we use the function *update.packages()* with the argument *ask = FALSE* which disallows prompting before updating each package. To delete a package already installed, we use the function *remove.packages()* by passing the name of the package to be removed as argument.

> *install.packages("chron")*

## 1.5. Environments and Functions

### 1.5.1. Environments

In R the variables that we create need to be stored in an environment. Environments are another type of variables. We can assign them, manipulate them and pass them as arguments to functions. They are like lists that are used to store different types of variables. When a variable is assigned in the command prompt, it goes by default into the *global environment*. When a function is called, an environment is automatically created to store the function-related variables. A new environment is created using the function *new.env()*.

> *newenvironment <- new.env()*

We can assign variables into a newly created environment using the *double square brackets* or the *dollar operator* as below.

> *newenvironment[["variable1"]] <- c(4, 7, 9)*

> *newenvironment$variable2 <- TRUE*

> *assign("variable3", "Value for variable3", newenvironment)*

The *assign()* function can also be used to assign variables to an environment. Retrieving values stored in an environment is like list indexing or we can use the *get()* function.

> *newenvironment[["variable1"]]*

*[1] 4 7 9*

> *newenvironment$variable2*

*[1] TRUE*

> *get("variable3", newenvironment)*

*[1] "Value for variable3"*

The functions *ls()* and *ls.str()* take an environment argument and lists its contents. We can test if a variable exists in an environment using the *exists()* function.

```
> ls(envir = newenvironment)
[1] "variable1" "variable2" "variable3"
> ls.str(envir = newenvironment)
variable1 : num [1:3] 4 7 9
variable2 : logi TRUE
variable3 : chr "Value for variable3"
> exists("variable2", newenvironment)
[1] TRUE
```

An environment can be converted into a list using the function *as.list()* and a list can be converted into an environment using the function *as.environment()* or the function *list2env()*.

```
> newlist <- as.list(newenvironment)
> newlist
$variable3
[1] "Value for variable3"
$variable1
[1] 4 7 9
$variable2
[1] TRUE
> as.environment(newlist)
<environment: 0x124730a8>
> list2env(newlist)
<environment: 0x12edf3e8>
> anotherenv <- as.environment(newlist)
> anotherenv[["variable3"]]
[1] "Value for variable3"
```

All environments are nested and so every environment has a parent environment. The empty environment sits at the top of the hierarchy without any parent. The *exists()* and the *get()* function also looks for the variables in the parent environment. To change this behaviour we need to pass the argument *inherits = FALSE*.

```
> subenv <- new.env(parent = newenvironment)
> exists("variable1", subenv)
[1] TRUE
> exists("variable1", subenv, inherits = FALSE)
[1] FALSE
```

The word frame is used interchangeably with the word environment. The function to refer to parent environment is denoted as *parent.frame()*. The variables assigned from the command prompt are stored in the *global* environment. The functions and the variables from the R's base package are stored in the *base* environment.

## 1.5.2. Functions

A function and its environment together is called a *closure*. When we load a package, the functions in that package are stored in the environment on the search path where the package is installed. A function in R is a verb and not a noun as it does things with its data. Functions are also another data types and hence we can assign and manipulate and pass them as arguments to other functions. Typing the function name in the command prompt lists the code associated with the function. Below is the code listed for the function *readLines()*.

```
> readLines
function (con = stdin(), n = -1L, ok = TRUE, warn = TRUE, encoding = unknown",
                                                      skipNul = FALSE)
{
    if (is.character(con)) {
        con <- file(con, "r")
        on.exit(close(con))
    }
    .Internal(readLines(con, n, ok, warn, encoding, skipNul))
}
```

When we call a function by passing values to it, the values are called as arguments. The lines of code of the function can be seen between the curly braces as body of the function. In R, there is no explicit return statement to return values. The last value that is calculated in a function is returned by default in R.

To create user defined functions, it is required to just assign the function as we do for other variables. The below code is an example of how to create a user defined function. In this *cube* is the name of the function and *x* is the argument passed to this function. The content within the curly braces is the body of the function. (*Note*: If it is a one line code we can omit the curly braces). Once a function is defined, it can be called like any other function in R by passing its arguments.

```
> cube <- function(x)
{ cu <- x ^ 3}
> z <- cube(5)
> z
[1] 125
```

The functions *formals()*, *args()* and *formalArgs()* can fetch the arguments defined for a function. The body of the function can be retrieved using the *body()* and *deparse()* functions.

```
> formals(cube)
$x
> args(cube)
function (x)
NULL
> formalArgs(cube)
[1] "x"
> body(cube)
{
    cu <- x^3
}
```

> *deparse(cube)*

[1] *"function (x) " "{"        "   cu <- x^3" "}"*

Functions can be passed as arguments to other functions and they can be returned from other functions. For calling a function, there is another function called *do.call()* in which we can pass the function name and its arguments as arguments. The use of this function can be seen below when using the *rbind()* function to concatenate two data frames.

> *f1 <- data.frame(x = 1:4, y = 5:8)*

> *f2 <- data.frame(x = 9:12, y = 13:16)*

> *do.call(rbind, list(f1, f2))*

|   | x | y |
|---|---|---|
| 1 | 1 | 5 |
| 2 | 2 | 6 |
| 3 | 3 | 7 |
| 4 | 4 | 8 |
| 5 | 9 | 13 |
| 6 | 10 | 14 |
| 7 | 11 | 15 |
| 8 | 12 | 16 |

When using functions as arguments to the *do.call()* function, it is not necessary to assign them first. We can pass a function anonymously as below.

> *do.call(function(x,y) x * y, list(1:3, 4:6))*

[1] *4 10 18*

## 1.5.3. Variable Scope

Variable's scope is the place where we can see the variable. If a variable is defined within a function, the variable can be accessed from any statement in the function. Also the sub-functions will have access to the variables defined in the parent function.

```
> x <- function(a1)
{
  a2 <- 1
  y <- function(a1)
  {
    a2 / a1
  }
  y(a1)
}
> x(5)
[1] 0.2
```

Thus R will search for a variable in the current environment and if it could not find it, it will check the same in its parent environment. This search will proceed upwards until the variable is searched in the global environment. The variables defined in the global environment are called the global variables, which can be accessed from anywhere else. The *replicate()* function can be used to run a function several times as below. In this the user defined function *random()* returns 1 if the value returned by the *rnorm()* function is a positive value and otherwise it returns the value of the argument passed to the function *random()*. This function *random()* is called 20 times using the *replicate()* function.

```
> random <- function(x)
+ {
+   if(rnorm(1) > 0)
+   {r <- 1}
+   else
+   {r <- x}
+ }
> replicate(20, random(5))
[1] 5 5 1 1 5 1 5 5 5 5 5 5 5 5 1 1 5 1 5 5
```

## 1.6. Flow Control

In some situations it may be required to execute some code only if a condition is satisfied.

### 1.6.1. If and Else Statement

The *if* statement takes a logical value and executes the next statement only if the value is TRUE.

> *> if(TRUE) message("TRUE Statement")*
>
> *TRUE Statement*
>
> *> if(FALSE) message("FALSE Statement")*

It is not necessary to pass the logical values as *TRUE* or *FALSE* directly, instead a variable or expression that returns a logical value can be used. If there are several statements to execute after the condition, they can be wrapped in curly braces.

```
a <- 5
if(a < 7)
{
    b <- a * 5
    c <- b * 3
    message("b is ", b)
    message("c is ", c)
}
b is 25
c is 75
```

In the *if* and *else* construct the code that follows the *if* statement is executed if the condition is TRUE and the code that follows the *else* statement is executed if the condition is FALSE. It is important to note that the *else* statement must occur on the same line as the closing curly brace of the *if* statement and otherwise it will throw an error message.

```
a <- 8
if(a < 7)
{
    b <- a * 5
    c <- b * 3
    message("b is ", b)
    message("c is ", c)
} else
{
    message("a is greater than 7")
}
a is greater than 7
```

The *if* and *else* statements can be used repeatedly to code multiple conditions and this respective actions. In this case it is important to note that the *if* and the *else* statements are separated and they are not one word as *ifelse*. The *ifelse* function is of different use which will be covered shortly.

```
a <- -8
if(a < 0)
{
    message("a is negative")
} else if(a == 0)
{
    message("a is zero")
} else if(a > 0)
{
    message("a is positive")

a is negative
```

The *ifelse()* function takes three arguments of which the first is logical condition, the second is the value that is returned when the first vector is TRUE and the third is the value that is returned when the first vector is FALSE.

```
> a <- 3
> b <- 5
> ifelse(a < b, "a is less than b", "a is greater than b")
[1] "a is less than b"
```

## 1.6.2. Switch Statement

If there are many else statements, it looks confusing and in such cases the *switch()* function is required. The first argument of the switch statement is an expression that can return a string value or an integer. This is followed by several named arguments that provide the results when the name matches the value of the first argument. Here also we can execute multiple statements enclosed by curly braces. If there is no match the switch statement returns *NULL*. So, in this case, it is safe to mention a default value if none matches.

```
> switch("color","color" = "red", "shape" = "circle", "radius" = 10)
[1] "red"
> switch("position","color" = "red", "shape" = "circle", "radius" = 10)
[1] NULL
> switch("position","color" = "red", "shape" = "circle", "radius" = 10,"default")
[1] "default"
> switch(2,"red","green","blue")
[1] "green"
```

## 1.7. Loops

There are three kinds of loops in R namely, *repeat, while* and *for.*

### 1.7.1. Repeat Loops

The *repeat* is the easiest loop in R that executes the same code until it is forced to stop. This *repeat* is similar to the *do while* statement in other languages. A *break* statement can be given when it is required to break the looping. Also, it is possible to skip the rest of the statements in a loop and execute the next iteration and this is done using the *next* statement.

```
a <- 1
repeat {
  message("Inside the loop")
  if(a == 3)
  {
      a = a + 1
      next
  }
  message("The value of a is ", a)
  a = a + 1
  if(a > 5)
  {
      message("Exiting the loop")
      break
  }
}
Inside the loop
The value of a is 1
Inside the loop
The value of a is 2
Inside the loop
Inside the loop
```

*The value of a is 4*

*Inside the loop*

*The value of a is 5*

*Exiting the loop*

## 1.7.2. While Loops

The *while* loops are backward *repeat* loops. The *repeat* loop executes the code and then checks for the condition, but in *while* loops the condition is first checked and then the code is executed. So, in this case it is possible that the code may not be executed even once when the condition fails at the entry itself during the first iteration. The same example above can be written using the while statement.

```
a <- 1
while (a <= 5)
{
    message("Inside the loop")
    if(a == 3)
    {
        a = a + 1
        next
    }
    message("The value of a is ", a)
    a = a + 1
}
```

*Inside the loop*

*The value of a is 1*

*Inside the loop*

*The value of a is 2*

*Inside the loop*

*Inside the loop*

*The value of a is 4*

*Inside the loop*

*The value of a is 5*

## 1.7.3. For Loops

The *for* loops are used when we know how many times the code needs to be repeated. The for loop accepts an iterating variable and a vector. It repeats the loop giving the iterating each element from the vector in turn. In this case also if there are multiple statements to execute, we can use the curly braces. The iterating variable can be an integer, number, character or logical vectors and they can be even lists.

*for(i in 1:5)*

*{*

   *j <- i \* i*

   *message("The square value of ", i, " is ", j)*

*}*

*The square value of 1 is 1*

*The square value of 2 is 4*

*The square value of 3 is 9*

*The square value of 4 is 16*

*The square value of 5 is 25*

*for(i in c(TRUE, FALSE, NA))*

*{*

   *message("This Statement is ", i)*

*}*

*This Statement is TRUE*

*This Statement is FALSE*

*This Statement is NA*

*a <- c(1,2,3)*

*b <- c("a","b","c","d")*

*d <- c(TRUE, FALSE)*

*l = list(a, b, d)*

*for(i in l)*

  *{*

  *message("The value of the list is ", i)*

  *}*

*The value of the list is 123*

*The value of the list is abcd*

*The value of the list is TRUEFALSE*

## ❖  HIGHLIGHTS

- R is a free open source language that has cross platform compatibility.
- R's syntax is very simple and intuitive.
- R's installation software can be downloaded from the CRAN Website.
- Help in R can be obtained by using, for eg. *?mean() / help("mean")*
- Variables can be assigned using the symbol ß or the *assign()* function.
- The basic functions are *c(), sum(), mean(), median(), exp(), sqrt()* etc.
- The basic operators are "+", "*", ":", "/", "**", "*", "%%", "%/%", "==", "!=", "<", ">", "<=", ">=" etc.
- Currently, the CRAN package repository features 10756 available packages.
- A Package can be newly installed using the function *install.packages()* and it can be invoked using the function *library()*.
- When a variable is assigned in the command prompt, it goes by default into the global environment.
- To create a new environment we use the function *new.env()*.
- Typing the function name in the command prompt lists the code associated with the function.
- The *if* and the *else* statements are separated and they are not one word as *ifelse*.
- The *ifelse()* function takes three arguments.
- If there are many else statements, the *switch()* function is required.

- The function *repeat* is similar to the *do while* statement in other languages.
- A *break* statement can be given when it is required to break the looping.
- To skip the rest of the statements in a loop we use the *next* statement.
- The *while* loops are backward *repeat* loops.
- The *for* loops are used when we know how many times the code needs to be repeated.

# CHAPTER 2

# DATA TYPES IN R

❖ **OBJECTIVES**

On completion of this Chapter you will be able to:

- know the basic data types in R of which the other complex data types are made of
- know how to create, access and perform basic operations on the vector data types in R
- know how to create, access and perform basic operations on matrices and arrays in R
- know how to create, access and perform basic operations on the list data types
- know how to create, access and perform basic operations on the factor data types in R
- know how to create, access and perform basic operations on strings in R
- understand the various date and time classes in R
- convert between various date formats
- setup various time zones
- perform calculations on dates and times

## 2.1. Basic Data Types in R

In contrast to other programming languages like C and Java, in R, the variables are not declared as some data type. The variables are assigned with R-Objects and the data type of the R-objects becomes the data type of the variables. There are many types of R-objects. The frequently used ones are — Vectors, Arrays, Matrices, Lists, Data Frames, Strings and Factors.

The simplest of these objects is the Vector object and there are six data types of these atomic vectors, also termed as six classes of vectors. The other R-Objects are built upon the atomic vectors. Hence, the basic data types in R are Numeric, Integer, Complex, Logical and Character.

## 2.1.1. Numeric

Decimal values are called numeric in R. It is the default computational data type. If we assign a decimal value to a variable x as follows, x will be of numeric type.

> *> x = 10.5*
>
> *> x*
>
> *[1] 10.5*
>
> *> class(x)      # print the class name of x*
>
> *[1] "numeric"*

Further more, even if we assign an integer to a variable *k*, it is still being saved as a numeric value. The fact that if *k* is an integer can be confirmed with the *is.integer()* function.

> *> k = 1*
>
> *> k*
>
> *[1] 1*
>
> *> class(k)*
>
> *[1] "numeric"*
>
> *> is.integer(k)*
>
> *[1] FALSE*

## 2.1.2. Integer

In order to create an integer variable in R, the *as.integer()* function is invoked as below.

```
> y = as.integer(3)
> y
[1] 3
> class(y)
[1] "integer"
> is.integer(y)
[1] TRUE
```

We can force a numeric value into an integer with the same *as.integer()* function as below.

```
> as.integer(3.14)
[1] 3
```

Similarly we can parse a string for a decimal value as below.

```
> as.integer("5.27")        # force a decimal string
[1] 5
```

But, if a non decimal string is forced, it is an error and it returns NA.

```
> as.integer("abc")
[1] NA
Warning message:
NAs introduced by coercion
```

The integer values of the logical values *TRUE* and *FALSE* are 1 and 0 respectively.

```
> as.integer(TRUE)
[1] 1
> as.integer(FALSE)
[1] 0
```

## 2.1.3. Complex

A complex number is expressed as an imaginary value *i*.

> *z = 3 + 4i*

> *z*

*[1] 3 + 4i*

> *class(z)*

*[1] "complex"*

If we find the square root of -1, it gives an error. But, if it is converted into a complex number and then square root is applied, it produces the necessary result as another complex number.

> *sqrt(-1)*

*[1] NaN*

*Warning message:*

*In sqrt(-1) : NaNs produced*

> *sqrt(as.complex(-1))*

*[1] 0+1i*

## 2.1.4. Logical

When two variables are compared, the logical values are created. The logical operators are "&" (and), "|" (or), and "!" (negation).

> *a = 4; b = 7*

> *p = a > b*

> *p*

*[1] FALSE*

> *class(p)*

*[1] "logical"*

> *a = TRUE; b = FALSE*

> *a & b*

*[1] FALSE*

> *a | b*

*[1] TRUE*

*> !a*

*[1] FALSE*

## 2.1.5. Character

The character object is used to represent string values in R. Objects can be converted into character values using the *as.character()* function. A *paste()* function can be used to concatenate two character values.

*> s = as.character("7.48")*

*> s*

*[1] "7.48"*

*> class(s)*

*[1] "character"*

*> fname = "Adam"*

*> lname = "Smith"*

*> paste(fname, lname)*

*[1] "Adam Smith"*

However, a readable string can be created using the *sprint()* function and this is similar to the C language syntax.

*> sprintf("%s has %d rupees", "Sundar",1000)*

*[1] "Sundar has 1000 rupees"*

The *substr()* function can be used to extract a substring from a given string. The *sub()* function is used to replace the first occurrence of a string with another string as below.

*> substr("Twinkle Twinkle Little Star", start = 9, stop = 15)*

*[1] "Twinkle"*

*> sub("Twinkle", "Wrinkle", "Twinkle Twinkle Little Star")*

*[1] "Wrinkle Twinkle Little Star"*

## 2.2. Vectors

A sequence of data elements of the same basic type is called a Vector. Members in a vector are called as components or members. The *vector()* function creates a vector of a specified type and length. The result is a zero or FALSE or empty string.

> *vector("numeric", 3)*

*[1] 0 0 0*

> *vector("logical", 5)*

*[1] FALSE FALSE FALSE FALSE FALSE*

> *vector("character", 2)*

*[1] "" ""*

The below commands also produces the same result as the above commands.

> *numeric(3)*

*[1] 0 0 0*

> *logical(5)*

*[1] FALSE FALSE FALSE FALSE FALSE*

> *character(2)*

*[1] "" ""*

The *seq()* function allows to generate sequences. The function *seq.int()* also creates sequence from one number to another, but this function provides more options for splitting the sequence.

> *seq(1:5)*

*[1] 1 2 3 4 5*

> *seq.int(5, 12)*

*[1] 5 6 7 8 9 10 11 12*

> *seq.int(10, 5, -1.5)*

*[1] 10.0 8.5 7.0 5.5*

The function *seq_len()* creates a sequence from 1 to the input value. The function *seq_along()* creates a sequence from 1 to the length of the input.

```
> seq_len(7)
[1] 1 2 3 4 5 6 7
> p <- c(3, 4, 5, 6)
> seq_along(p)
[1] 1 2 3 4
```

The function *length()* can be used to find the length of the vector, that is the number of elements in a vector. Using this function, it is possible to assign new length to a vector. If the vector length is extended NA(s) will be added to the end.

```
> length(1:7)
[1] 7
> length(c("aa", "ccc", "eeee"))
[1] 3
> nchar(c("aa", "ccc", "eeee"))
[1] 2 3 4
> s <- c(1,2,3,4,5)
> length(s) <- 3
> s
[1] 1 2 3
> length(s) <- 8
> s
[1] 1 2 3 NA NA NA NA NA
```

Each element of a vector can be given a name during the vector creation itself. If there are space or special characters in the name, it needs to be enclosed in quotes. The *names()* function can be used to give names to the vector elements after its creation.

```
> c(a = 1, b = 2, c = 3)
a b c
1 2 3
> s <- 1:3
> s
[1] 1 2 3
> names(s) <- c("a", "b", "c")
> s
a b c
1 2 3
```

Elements of a vector can be accessed using its indexes which are specified in a square bracket. The index number starts from 1 and not 0. Specifying a negative number as index to a vector means, it returns all the elements except the one specified. The name of the vector element can also be specified as index to fetch it.

```
> x <- c(1:5)
> x
[1] 1 2 3 4 5
> x[c(2,3)]
[1] 2 3
> x[c(-1,-4)]
[1] 2 3 5
> s <- 1:3
> s
[1] 1 2 3
> names(s) <- c("a", "b", "c")
> s["b"]
b
2
```

If an incorrect index is specified to access a vector element, the result is NA. Non integer indices are rounded off. Not passing any index to a vector will return all the elements of the vector.

```
> x
[1] 1 2 3 4 5
> x[7]
[1] NA
```

The *which()* function returns the elements of the vector which satisfies the condition specified within this function. The functions *which.min()* and *which.max()* can be used to display the minimum and the maximum elements in the vector.

```
> x
[1] 1 2 3 4 5
> which.min(x)
[1] 1
> which.max(x)
[1] 5
> which(x>3)
[1] 4 5
```

Vectors can be combined using the *c()* function. When the two vectors are combined the numeric values are forced into character values. This shows that all the members of a vector should be of the same basic data type.

```
> f = c(7, 5, 9)
> g = c("aaa", "bbb", "ccc")
> c(f, g)
[1] "7"  "5"  "9"  "aaa" "bbb" "ccc"
```

Arithmetic operations in a vector will be performed member-wise. If two vectors are of unequal length, the shorter vector will be recycled in order to match the longer vector.

```
> x = c(5, 8, 9)
> y = c(2, 6, 9)
> 4 * y
[1]  8 24 36
> x + y
[1]  7 14 18
> x - y
[1] 3 2 0
> x * y
[1] 10 48 81
> x / y
[1] 2.500000 1.333333 1.000000
> v = c(1, 2, 3, 4, 5, 6)
> x + v
[1]  6 10 12  9 13 15
```

The *rep()* function creates a vector with repeated elements. This function has its other variants such as *rep.int()* and *rep_len()* whose usage is as given below.

```
> rep(1:3, 4)
 [1] 1 2 3 1 2 3 1 2 3 1 2 3
> rep(1:3, each = 4)
 [1] 1 1 1 1 2 2 2 2 3 3 3 3
> rep(1:3, times = 1:3)
[1] 1 2 2 3 3 3
> rep(1:3, length.out = 9)
[1] 1 2 3 1 2 3 1 2 3
> rep.int(1:3, 4)
 [1] 1 2 3 1 2 3 1 2 3 1 2 3
```

> *rep_len(1:3, 9)*

*[1] 1 2 3 1 2 3 1 2 3*

## 2.3. Matrices and Arrays

A matrix is a collection of data elements with the same basic type arranged in a two-dimensional rectangular layout. An array consists of multidimensional rectangular data. Matrices are special cases of two-dimensional arrays. To create an array the *array()* function can be used and a vector of values and vector of dimensions are passed to it.

> *x <- array(1:24, dim = c(4, 3, 2),*

>          *dimnames = list(c("a", "b", "c", "d"), c("e", "f", "g"), c("h", "i")))*

> *x*

*, , h*

   *e f g*

*a 1 5 9*

*b 2 6 10*

*c 3 7 11*

*d 4 8 12*

*, , i*

   *e f g*

*a 13 17 21*

*b 14 18 22*

*c 15 19 23*

*d 16 20 24*

The syntax for creating matrices is using the function *matrix()* and passing the *nrow* or *ncol* argument instead of the *dim* argument in the arrays. A matrix can also be created using the *array()* function where the dimension of the array is two.

> *m <- matrix(1:12, nrow = 3, dimnames = list(c("a", "b", "c"), c("d", "e", "f", "g")))*

> *m*

 *d e f g*

*a 1 4 7 10*

*b 2 5 8 11*

*c 3 6 9 12*

*> m1 <- array(1:12, dim = c(3,4),*

*dimnames = list(c("a", "b", "c"), c("d", "e", "f", "g")))*

*> m1*

 *d e f g*

*a 1 4 7 10*

*b 2 5 8 11*

*c 3 6 9 12*

The argument *byrow = TRUE* in the *matrix()* function assigns the elements row wise. If this argument is not specified, by default the elements are filled column wise.

*> m <- matrix(1:12, nrow = 3, byrow = TRUE,*

      *dimnames = list(c("a", "b", "c"), c("d", "e", "f", "g")))*

The *dim()* function returns the dimensions of an array or a matrix. The functions *nrow()* and *ncol()* returns the number of rows and number of columns of a matrix respectively.

*> dim(x)*

*[1] 4 3 2*

*> dim(m)*

*[1] 3 4*

*> nrow(m)*

*[1] 3*

*> ncol(m)*

*[1] 4*

The *length()* function also works for matrices and arrays. It is also possible to assign new dimension for a matrix or an array using the *dim()* function.

> *length(x)*

*[1] 24*

> *length(m)*

*[1] 12*

> *dim(m) <- c(6,2)*

The functions *rownames()*, *colnames()* and *dimnames()* can be used to fetch the row names, column names and dimension names of matrices and arrays respectively.

> *rownames(m1)*

*[1] "a" "b" "c"*

> *colnames(m1)*

*[1] "d" "e" "f" "g"*

> *dimnames(x)*

*[[1]]*

*[1] "a" "b" "c" "d"*

*[[2]]*

*[1] "e" "f" "g"*

*[[3]]*

*[1] "h" "i"*

It is possible to extract the element at the $n^{th}$ row and $m^{th}$ column using the expression    *M[n, m]*. The entire $n^{th}$ row can be extracted using *M[n, ]* and similarly, the $m^{th}$ column can be extracted using *M[,m]*. Also, it is possible to extract more than one column or row.

> *M[2,3]*

*[1] 6*

> *M[2,]*

*[1] 4 5 6*

```
> M[,3]
[1] 3 6 9
> M[,c(1,3)]
   [,1] [,2]
[1,]  1   3
[2,]  4   6
[3,]  7   9
> M[c(1,3),]
   [,1] [,2] [,3]
[1,]  1   2   3
[2,]  7   8   9
```

The matrix transpose is constructed by interchanging its rows and columns using the function $t()$.

```
> t(M)
   r1 r2 r3
c1  1  4  7
c2  2  5  8
c3  3  6  9
```

The columns of two matrices can be combined using the *cbind()* function and similarly the rows of two matrices can be combined using the *rbind()* function.

```
> M1 = matrix(c(2,4,6,8,10,12), nrow=3, ncol=2)
> M1
   [,1] [,2]
[1,]  2   8
[2,]  4  10
[3,]  6  12
> M2 = matrix(c(3,6,9), nrow=3, ncol = 1)
> M2
```

```
   [,1]
[1,]   3
[2,]   6
[3,]   9
> cbind(M1, M2)
   [,1] [,2] [,3]
[1,]   2   8   3
[2,]   4  10   6
[3,]   6  12   9
> M3 = matrix(c(4,8), nrow=1, ncol=2)
> M3
   [,1] [,2]
[1,]   4   8
> rbind(M1, M3)
   [,1] [,2]
[1,]   2   8
[2,]   4  10
[3,]   6  12
[4,]   4   8
```

A matrix can be deconstructed using the *c()* function which combines all column vectors into one.

```
> c(M1)
[1] 2 4 6 8 10 12
```

The arithmetic operators "+", "- ", "* ", "/ " work element wise on matrices and arrays. But the condition is that the matrices or arrays should be of conformable sizes. The matrix multiplication is done using the operator "%*%".

```
> M1 = matrix(c(2,4,6,8,10,12), nrow=3, ncol=2)
> M1
```

```
      [,1] [,2]
[1,]    2   8
[2,]    4  10
[3,]    6  12
> M2 = matrix(c(3,6,9,11,1,5), nrow=3, ncol = 2)
> M2
     [,1] [,2]
[1,]    3  11
[2,]    6   1
[3,]    9   5
> M1 + M2
     [,1] [,2]
[1,]    5  19
[2,]   10  11
[3,]   15  17
> M1 * M2
     [,1] [,2]
[1,]    6  88
[2,]   24  10
[3,]   54  60
> M2 = matrix(c(3,6,9,11), nrow=2, ncol = 2)
> M2
     [,1] [,2]
[1,]    3   9
[2,]    6  11
> M1 %*% M2
     [,1] [,2]
[1,]   54 106
```

*[2,]   72  146*

*[3,]   90  186*

The power operator "^" also works element wise on matrices. To find the inverse of a matrix the function *solve()* can be used.

> *M2*

  *[,1] [,2]*

*[1,]   3   9*

*[2,]   6   11*

> *M2 ^ -1*

  *[,1]     [,2]*

*[1,] 0.3333333 0.11111111*

*[2,] 0.1666667 0.09090909*

> *solve(M2)*

  *[,1]     [,2]*

*[1,] -0.5238095  0.4285714*

*[2,]  0.2857143 -0.1428571*

## 2.4. Lists

Lists allow us to combine different data types in a single variable. Lists can be created using the *list()* function. This function is similar to the *c()* function. The contents of a list are just listed within the *list()* function as arguments separated by a comma. List elements can be a vector, matrix or a function. It is possible to name the elements of the list while creation or later using the *names()* function.

> *L <- list(c(9,1, 4, 7, 0), matrix(c(1,2,3,4,5,6), nrow = 3))*

> *L*

*[[1]]*

*[1] 9 1 4 7 0*

*[[2]]*

```
      [,1] [,2]
[1,]   1    4
[2,]   2    5
[3,]   3    6

> names(L) <- c("Num", "Mat")

> L
$Num
[1] 9 1 4 7 0
$Mat
      [,1] [,2]
[1,]   1    4
[2,]   2    5
[3,]   3    6
> L <- list(Num = c(9,1, 4, 7, 0), Mat = matrix(c(1,2,3,4,5,6), nrow = 3))
> L
$Num
[1] 9 1 4 7 0
$Mat
      [,1] [,2]
[1,]   1    4
[2,]   2    5
[3,]   3    6
```

Lists can be nested. That is a list can be an element of another list. But, vectors, arrays and matrices are not recursive/nested. They are atomic. The functions *is.recursive()* and *is.atomic()* shows if a variable type is recursive or atomic respectively.

```
> is.atomic(list())
[1] FALSE
```

```
> is.recursive(list())
```
*[1] TRUE*
```
> is.atomic(L)
```
*[1] FALSE*
```
> is.recursive(L)
```
*[1] TRUE*
```
> is.atomic(matrix())
```
*[1] TRUE*
```
> is.recursive(matrix())
```
*[1] FALSE*

The *length()* function works on list like in vectors and matrices. But, the *dim()*, *nrow()* and *ncol()* functions returns only *NULL*.

```
> length(L)
```
*[1] 2*
```
> dim(L)
```
*NULL*
```
> nrow(L)
```
*NULL*
```
> ncol(L)
```
*NULL*

Arithmetic operations in list are possible only if the elements of the list are of the same data type. Generally, it is not recommended. As in vectors the elements of the list can be accessed by indexing them using the square brackets. The index can be a positive number, or a negative number, or element names or logical values.

```
> L1 <- list(l1 = c(8, 9, 1), l2 = matrix(c(1,2,3,4), nrow = 2),
                    l3 = list( l31 = c("a", "b"), l32 = c(TRUE, FALSE) ))
> L1
$l1
```

*[1] 8 9 1*

*$l2*

  *[,1] [,2]*

*[1,]  1   3*

*[2,]  2   4*

*$l3*

*$l3$l31*

*[1] "a" "b"*

*$l3$l32*

*[1]  TRUE FALSE*

*> L1[1:2]*

*$l1*

*[1] 8 9 1*

*$l2*

  *[,1] [,2]*

*[1,]  1   3*

*[2,]  2   4*

*> L1[-3]*

*$l1*

*[1] 8 9 1*

*$l2*

  *[,1] [,2]*

*[1,]  1   3*

*[2,]  2   4*

*> L1[c("l1", "l2")]*

*$l1*

*[1] 8 9 1*

*$l2*

  *[,1] [,2]*

*[1,]*  *1*   *3*

*[2,]*  *2*   *4*

*> L1[c(TRUE, TRUE, FALSE)]*

*$l1*

*[1] 8 9 1*

*$l2*

  *[,1] [,2]*

*[1,]*  *1*   *3*

*[2,]*  *2*   *4*

A list is a generic vector containing other objects.

*> a = c(4,8,12)*

*> b = c("abc", "def", "ghi", "jkl", "mno")*

*> d = c(TRUE, FALSE)*

*> t = list(a, b, d, 5)*

The list *t* contains copies of the vectors a, b and d. A list slice is retrieved using single square brackets *[]*. In the below, *t[2]* contains a slice and a copy of *b*. Slice can also be retrieved with multiple members.

*> t[2]*

*[[1]]*

*[1] "abc" "def" "ghi" "jkl" "mno"*

*> t[c(2,4)]*

*[[1]]*

*[1] "abc" "def" "ghi" "jkl" "mno"*

*[[2]]*

*[1] 5*

To reference a list member directly double square bracket *[[]]* is used. Thus *t[[2]]* retrieves the second member of the list *t*. This results in a copy of *b*, but not a slice of *b*. It is also possible to modify the contents of the elements directly, but the contents of *b* are unaffected.

> *t[[2]]*

*[1] "abc" "def" "ghi" "jkl" "mno"*

> *t[[2]][1] = "qqq"*

> *t[[2]]*

*[1] "qqq" "def" "ghi" "jkl" "mno"*

> *b*

*[1] "abc" "def" "ghi" "jkl" "mno"*

We can assign names to the list members and reference lists by names instead of numeric indexes. A list of two members is given as example below with the member names as *"first"* and *"second"*. The list slice containing the member *"first"* can be retrieved using the square brackets *[]* as shown below.

> *l = list(first=c(1,2,3), second=c("a","b", "c"))*

> *l*

*$first*

*[1] 1 2 3*

*$second*

*[1] "a" "b" "c"*

> *l["first"]*

*$first*

*[1] 1 2 3*

The named list member can also be directly referenced with the $ operator or double square brackets *[[]]* as below.

> *l$first*

*[1] 1 2 3*

> l[["first"]]

[1] 1 2 3

A vector can be converted to a list using the function *as.list()*. Similarly, a list can be converted into a vector, provided the list contains scalar elements of the same type. This is done using the conversion functions such as *as.numeric()*, *as.character()* and so on. If a list consists of non-scalar elements, but if they are of the same type, then it can be converted into a vector using the function *unlist()*.

> v <- c(7, 3, 9, 2, 6)

> as.list(v)

[[1]]

[1] 7

[[2]]

[1] 3

[[3]]

[1] 9

[[4]]

[1] 2

[[5]]

[1] 6

> L <- list(3, 7, 8, 12, 14)

> as.numeric(L)

[1]  3  7  8 12 14

> L1 <- list("aaa", "bbb", "ccc")

> L1

[[1]]

[1] "aaa"

[[2]]

[1] "bbb"

*[[3]]*

*[1] "ccc"*

*> as.character(L1)*

*[1] "aaa" "bbb" "ccc"*

*> L1 <- list(l1 = c(78, 90, 21), l2 = c(11,22,33,44,55))*

*> L1*

*$l1*

*[1] 78 90 21*

*$l2*

*[1] 11 22 33 44 55*

*> unlist(L1)*

*l11 l12 l13 l21 l22 l23 l24 l25*

 *78  90  21  11  22  33  44  55*

The *c()* function can also be used to combine lists as we do for vectors.

*> L1 <- list(l1 = c(78, 90, 21), l2 = c(11,22,33,44,55))*

*> L2 <- list("aaa", "bbb", "ccc")*

*> c(L1, L2)*

*$l1*

*[1] 78 90 21*

*$l2*

*[1] 11 22 33 44 55*

*[[3]]*

*[1] "aaa"*

*[[4]]*

*[1] "bbb"*

*[[5]]*

*[1] "ccc"*

## 2.5. Data Frames

A data frame is used for storing data tables. They store spread-sheet like data. It is a list of vectors of equal length (not necessarily of the same basic data type). Consider a data frame *df1* consisting of three vectors a, b, and d.

```
> a = c(1, 2, 3)
> b = c("a", "b", "c")
> d = c(TRUE, FALSE, TRUE)
> df1 = data.frame(a, b, d)
> df1
```

|   | a | b | d |
|---|---|---|---|
| 1 | 1 | a | TRUE |
| 2 | 2 | b | FALSE |
| 3 | 3 | c | TRUE |

By default the row names are automatically numbered from 1 to the number of rows in the data frame. It is also possible to provide row names manually using the row.names argument as below.

```
> df1 = data.frame(a, b, d, row.names = c("one", "two", "three"))
> df1
```

|       | a | b | d |
|-------|---|---|---|
| one   | 1 | a | TRUE |
| two   | 2 | b | FALSE |
| three | 3 | c | TRUE |

The functions rownames(), colnames(), dimnames(), nrow(), ncol() and dim() can be applied on the data frames as below. The length() and names() function, returns the same result as that of ncol() and colnames() respectively.

```
> rownames(df1)
[1] "one" "two" "three"
> colnames(df1)
```

```
[1] "a" "b" "d"
> dimnames(df1)
[[1]]
[1] "one"  "two"  "three"
[[2]]
[1] "a" "b" "d"

> nrow(df1)
[1] 3
> ncol(df1)
[1] 3
> dim(df1)
[1] 3 3
> length(df1)
[1] 3
> colnames(df1)
[1] "a" "b" "d"
```

It is possible to create data frames with different length of vectors as long as the shorter ones can be recycled to match that of the longer ones. Otherwise, an error will be thrown.

```
> df2 <- data.frame(x = 1, y = 2:3, y = 4:7)
> df2
```

|   | x | y | y.1 |
|---|---|---|-----|
| 1 | 1 | 2 | 4   |
| 2 | 1 | 3 | 5   |
| 3 | 1 | 2 | 6   |
| 4 | 1 | 3 | 7   |

The argument check.names can be set as FALSE so that a data frame will not look for valid column names.

> *df3 <- data.frame("BaD col" = c(1:5), "!@#$% ^ &*" = c("aaa"))*

> *df3*

   *BaD.col X........*

   *1    1    aaa*

   *2    2    aaa*

   *3    3    aaa*

   *4    4    aaa*

   *5    5    aaa*

There are many built-in data frames available in R (example – mtcars). When this data frame is invoked in R tool, it produces the below result.

> *mtcars*

| | mpg | cyl | disp | hp | drat | wt ... | |
|---|---|---|---|---|---|---|---|
| Mazda RX4 | 21.0 | 6 | 160 | 110 | 3.90 | 2.62 | ... |
| Mazda RX4 Wag | 21.0 | 6 | 160 | 110 | 3.90 | 2.88 | ... |
| Datsun 710 | 22.8 | 4 | 108 | 93 | 3.85 | 2.32 | ... |

   *............*

The top line contains the header or the column names. Each row denotes a record or a row in the table. A row begins with the name of the row. Each data member of a row is called a cell. To retrieve a cell value, we enter the row and the column number of the cell in square brackets [] separated by a comma. The cell value of the second row and third column is retrieved as below. The row and the column names can also be used inside the square brackets [] instead of the row and column numbers.

> *mtcars[2, 3]*

*[1] 160*

> *mtcars["Mazda RX4 Wag", "disp"]*

*[1] 160*

The *nrow()* function gives the number of rows in a data frame and the *ncol()* function gives the number of columns in a data frame. To get the preview or the first few records of a data frame along with the header the *head()* function can be used.

> *nrow(mtcars)*

*[1] 32*

> *ncol(mtcars)*

*[1] 11*

> *head(mtcars)*

|  | *mpg* | *cyl* | *disp* | *hp* | *drat* | *wt* | ... |
|---|---|---|---|---|---|---|---|
| *Mazda RX4* | *21.0* | *6* | *160* | *110* | *3.90* | *2.62 ...* | |

*......*

To retrieve a column from a data frame we use double square brackets *[[]]* and the column name or the column number inside the *[[]]*. The same can be achieved by making use of the $ symbol as well. This same result can also be achieved by using single brackets *[]* by mentioning a comma instead of the row name / number and using the column name / number as the second index inside the *[]*.

> *mtcars[["hp"]]*

*[1] 110 110 93 110 175 105 245 62 95 123 123 180 180 180 ....*

> *mtcars[[4]]*

*[1] 110 110 93 110 175 105 245 62 95 123 123 180 180 180 ....*

> *mtcars$hp*

*[1] 110 110 93 110 175 105 245 62 95 123 123 180 180 180 ....*

> *mtcars[,"hp"]*

*[1] 110 110 93 110 175 105 245 62 95 123 123 180 180 180 ....*

> *mtcars[,4]*

*[1] 110 110 93 110 175 105 245 62 95 123 123 180 180 180 ....*

Similarly, if we use the column name or the column number inside a single square bracket *[]*, we get the below result.

> *mtcars[4]*

|  | *hp* |
|---|---|
| *Mazda RX4* | *110* |

| | | |
|---|---|---|
| *Mazda RX4 Wag* | | *110* |
| *Datsun 710* | | *93* |

....

> *mtcars[c("mpg","hp")]*

| | *mpg* | *hp* |
|---|---|---|
| *Mazda RX4* | *21.0* | *110* |
| *Mazda RX4 Wag* | *21.0* | *110* |
| *Datsun 710* | *22.8* | *93* |

....

To retrieve a row from a data frame we use the single square brackets [] only by mentioning the row name / number as the first index inside [] and a comma instead of the column name / number.

> *mtcars[6,]*

| | *mpg* | *cyl* | *disp* | *hp* | *drat* | *wt....* |
|---|---|---|---|---|---|---|
| *Valiant* | *18.1* | *6* | *225* | *105* | *2.76* | *3.46....* |

> *mtcars[c(6,18),]*

| | *mpg* | *cyl* | *disp* | *hp* | *drat* | *wt....* |
|---|---|---|---|---|---|---|
| *Valiant* | *18.1* | *6* | *225* | *105* | *2.76* | *3.46....* |
| *Fiat 128* | *32.4* | *4* | *78.7* | *66* | *4.08* | *2.20....* |

> *mtcars["Valiant",]*

| | *mpg* | *cyl* | *disp* | *hp* | *drat* | *wt....* |
|---|---|---|---|---|---|---|
| *Valiant* | *18.1* | *6* | *225* | *105* | *2.76* | *3.46....* |

> *mtcars[c("Valiant","Fiat 128"),]*

| | *mpg* | *cyl* | *disp* | *hp* | *drat* | *wt....* |
|---|---|---|---|---|---|---|
| *Valiant* | *18.1* | *6* | *225* | *105* | *2.76* | *3.46....* |
| *Fiat 128* | *32.4* | *4* | *78.7* | *66* | *4.08* | *2.20....* |

If we need to fetch a subset of a data frame by selecting few columns and specifying conditions on the rows, we can use the subset() function to do this. This function takes the arguments, the data frame, the condition to be applied on the rows and the columns to be fetched.

> *x <- c("a", "b", "c", "d", "e", "f")*

> *y <- c(3, 4, 7, 8, 12, 15)*

> *z <- c(TRUE, TRUE, FALSE, TRUE, FALSE, TRUE)*

> *D <- data.frame(x, y, z)*

> *D*

|   | *x* | *y* | *z* |
|---|-----|-----|-----|
| 1 | *a* | 3 | TRUE |
| 2 | *b* | 4 | TRUE |
| 3 | *c* | 7 | FALSE |
| 4 | *d* | 8 | TRUE |
| 5 | *e* | 12 | FALSE |
| 6 | *f* | 15 | TRUE |

> *subset(D, y<10 & z, x)*

|   | *x* |
|---|-----|
| 1 | *a* |
| 2 | *b* |
| 4 | *d* |

As we have for matrices the transpose of a data frame can be obtained using the t() function as below.

> *t(D)*

|   | *[,1]* | *[,2]* | *[,3]* | *[,4]* | *[,5]* | *[,6]* |
|---|--------|--------|--------|--------|--------|--------|
| *x* | "a" | "b" | "c" | "d" | "e" | "f" |
| *y* | " 3" | " 4" | " 7" | " 8" | "12" | "15" |
| *z* | " TRUE" | " TRUE" | "FALSE" | " TRUE" | "FALSE" | " TRUE" |

The functions *rbind()* and *cbind()* can also be applied on the data frames as we do for the matrices. The only condition for *rbind()* is that the column names should match, but for *cbind()* it does not check even if the column names are duplicated.

```
> x1 <- c("aaa", "bbb", "ccc", "ddd", "eee", "fff")
> y1 <- c(9, 12, 17, 18, 23, 32)
> z1 <- c(TRUE, FALSE, TRUE, FALSE, TRUE, FALSE)
> E <- data.frame(x1, y1, z1)
> E
```

|   | x1 | y1 | z1 |
|---|-----|-----|-------|
| 1 | aaa | 9 | TRUE |
| 2 | bbb | 12 | FALSE |
| 3 | ccc | 17 | TRUE |
| 4 | ddd | 18 | FALSE |
| 5 | eee | 23 | TRUE |
| 6 | fff | 32 | FALSE |

```
> cbind(D, E)
```

|   | x | y | z | x1 | y1 | z1 |
|---|---|----|-------|-----|-----|-------|
| 1 | a | 3 | TRUE | aaa | 9 | TRUE |
| 2 | b | 4 | TRUE | bbb | 12 | FALSE |
| 3 | c | 7 | FALSE | ccc | 17 | TRUE |
| 4 | d | 8 | TRUE | ddd | 18 | FALSE |
| 5 | e | 12 | FALSE | eee | 23 | TRUE |
| 6 | f | 15 | TRUE | fff | 32 | FALSE |

```
> F <- data.frame(x, y, z)
> F
```

|    | *x* | *y* | *z* |
|----|-----|-----|-----|
| 1  | a   | 9   | TRUE  |
| 2  | b   | 12  | FALSE |
| 3  | c   | 17  | TRUE  |
| 4  | d   | 18  | FALSE |
| 5  | e   | 23  | TRUE  |
| 6  | f   | 32  | FALSE |

> *rbind(D, F)*

|    | *x* | *y* | *z* |
|----|-----|-----|-----|
| 1  | a   | 3   | TRUE  |
| 2  | b   | 4   | TRUE  |
| 3  | c   | 7   | FALSE |
| 4  | d   | 8   | TRUE  |
| 5  | e   | 12  | FALSE |
| 6  | f   | 15  | TRUE  |
| 7  | a   | 9   | TRUE  |
| 8  | b   | 12  | FALSE |
| 9  | c   | 17  | TRUE  |
| 10 | d   | 18  | FALSE |
| 11 | e   | 23  | TRUE  |
| 12 | f   | 32  | FALSE |

The merge() function can be applied to merge two data frames provided they have common column names. By default, the merge() function does the merging based on all the common columns, otherwise one of the common column name has to be specified.

> *merge(D, F, by = "x", all = TRUE)*

|   | x | y.x | z.x | y.y | z.y |
|---|---|-----|-------|-----|-------|
| 1 | a | 3 | TRUE | 9 | TRUE |
| 2 | b | 4 | TRUE | 12 | FALSE |
| 3 | c | 7 | FALSE | 17 | TRUE |
| 4 | d | 8 | TRUE | 18 | FALSE |
| 5 | e | 12 | FALSE | 23 | TRUE |
| 6 | f | 15 | TRUE | 32 | FALSE |

The functions colSums(), colMeans(), rowSums() and rowMeans() can be applied on the data frames that have numeric values as below.

> x <- c(5, 6, 7, 8)
> y <- c(15, 16, 17, 18)
> z <- c(25, 26, 27, 28)
> G <- data.frame(x, y, z)
> G

|   | x | y | z |
|---|---|---|---|
| 1 | 5 | 15 | 25 |
| 2 | 6 | 16 | 26 |
| 3 | 7 | 17 | 27 |
| 4 | 8 | 18 | 28 |

> colSums(G[, 1:2])

| x | y |
|----|----|
| 26 | 66 |

> colMeans(G[, 1:3])

| x | y | z |
|-----|------|------|
| 6.5 | 16.5 | 26.5 |

> *rowSums(G[1:3, ])*

|  1  |  2  |  3  |
|-----|-----|-----|
| 45  | 48  | 51  |

> *rowMeans(G[2:4, ])*

|  2  |  3  |  4  |
|-----|-----|-----|
| 16  | 17  | 18  |

## 2.6. Factors

Factors are used to store categorical data like gender ("Male" or "Female"). They behave sometimes like character vectors and sometimes like integer vectors based on the context.

Factors stores categorical data and they behave like strings sometimes and integers sometimes. Consider a data frame that stores the weight of few males and females. In this case the column that stores the gender is a factor as it stores categorical data. The choices "female" and "male" are called the levels of the factor. This can be viewed by using the *levels()* function and *nlevels()* function.

> *weight <- data.frame(wt_kg = c(60,82,45, 49,52,75,68),*

   *gender = c("female","male", "female", "female", "female", "male", "male"))*

> *weight*

|   | wt_kg | gender |
|---|-------|--------|
| 1 | 60    | female |
| 2 | 82    | male   |
| 3 | 45    | female |
| 4 | 49    | female |
| 5 | 52    | female |
| 6 | 75    | male   |
| 7 | 68    | male   |

> *weight$gender*

*[1] female male   female female female male   male*

*Levels: female male*

*> levels(weight$gender)*

*[1] "female" "male"*

*> nlevels(weight$gender)*

*[1] 2*

At the atomic level a factor can be created using the *factor()* function, which takes a character vector as the argument.

*> gender <- factor(c("female", "male", "female", "female", "female", "male", "male"))*

*> gender*

*[1] female male   female female female male   male*

*Levels: female male*

The levels argument can be used in the *factor()* function to specify the levels of the factor. It is also possible to change the levels once the factor is created. This is done using the function *levels()* or the function *relevel()*. The function *relevel()* just mentions which level comes first.

*> gender <- factor(c("female", "male", "female", "female", "female",*

*"male", "male"), levels = c("male", "female"))*

*> gender*

*[1] female male   female female female male   male*

*Levels: male female*

*> levels(gender) <- c("F", "M")*

*> gender*

*[1] M F M M M F F*

*Levels: F M*

*> relevel(gender, "M")*

*[1] M F M M M F F*

*Levels: M F*

It is possible to drop a level from a factor using the function *droplevels()* when the level is not in use as in the example below. [*Note:* the function *is.na()* is used to remove the missing value].

> diet <- data.frame(eat = c("fruit", "fruit", "vegetable", "fruit"),

type = c("apple", "mango", NA, "papaya"))

> diet

|   | eat | type |
|---|---|---|
| 1 | fruit | apple |
| 2 | fruit | mango |
| 3 | vegetable | <NA> |
| 4 | fruit | papaya |

> diet <- subset(diet, !is.na(type))

> diet

|   | eat | type |
|---|---|---|
| 1 | fruit | apple |
| 2 | fruit | mango |
| 4 | fruit | papaya |

> diet$eat

[1] fruit fruit fruit

Levels: fruit vegetable

> levels(diet)

NULL

> levels(diet$eat)

[1] "fruit"   "vegetable"

> unique(diet$eat)

[1] fruit

Levels: fruit vegetable

```
> diet$eat <- droplevels(diet$eat)
> levels(diet$eat)
[1] "fruit"
```

In some cases, the levels need to be ordered as in rating a product or course. The ratings can be *"Outstanding"*, *"Excellent"*, *"Very Good"*, *"Good"*, *"Bad"*. When a factor is created with these levels, it is not necessary they are ordered. So, to order the levels in a factor, we can either use the function *ordered()* or the argument *ordered = TRUE* in the *factor()* function. Such ordering can be useful when analysing survey data.

```
> ch <- c("Outstanding", "Excellent", "Very Good", "Good", "Bad")
> val <- sample(ch, 100, replace = TRUE)
> rating <- factor(val, ch)
> rating
  [1] Outstanding Bad       Outstanding Good      Very Good  Very Good
  [7] Excellent   Outstanding Bad       Excellent  Very Good  Bad
...
Levels: Outstanding Excellent Very Good Good Bad
> is.factor(rating)
[1] TRUE
> is.ordered(rating)
[1] FALSE
> rating_ord <- ordered(val, ch)
> is.factor(rating_ord)
[1] TRUE
> is.ordered(rating_ord)
[1] TRUE
> rating_ord
```

*[1] Outstanding Bad      Outstanding Good      Very Good   Very Good*

*[7] Excellent   Outstanding Bad        Excellent   Very Good   Bad*

...

*Levels: Outstanding < Excellent < Very Good < Good < Bad*

Numeric values can be summarized into factors using the *cut()* function and the result can be viewed using the *table()* function which lists the count of numbers in each category. For example let us consider the variable *age* which has the numeric values of ages. These ages can be grouped using the *cut()* function with an interval of 10 and the result is a factor *age_group*.

```
> age <- c(18,20, 31, 32, 33, 35, 41, 38, 45, 48, 51, 27, 29, 42, 39)

> age_group <- cut(age, seq.int(15, 55, 10))

> age

[1] 18 20 31 32 33 35 41 38 45 48 51 27 29 42 39

> age_group

[1] (15,25] (15,25] (25,35] (25,35] (25,35] (25,35] (35,45] (35,45] (35,45] (45,55]

[11] (45,55] (25,35] (25,35] (35,45] (35,45]

Levels: (15,25] (25,35] (35,45] (45,55]

> table(age_group)

age_group

(15,25] (25,35] (35,45] (45,55]

   2      6      5     2
```

The function *gl()* can be used to create a factor, which takes the first argument that tells how many levels the factor contains and the second argument that tells how many times each level has to be repeated as value. This function can also take the argument *labels*, which lists the names of the factor levels. The function can also be made to list alternating values of the labels as below.

```
> gl(5,3)

[1] 1 1 1 2 2 2 3 3 3 4 4 4 5 5 5

Levels: 1 2 3 4 5
```

> *gl(5,3, labels = c("one", "two", "three", "four", "five"))*

*[1] one  one  one  two  two  two  three three three four  four  four  five*

*[14] five  five*

*Levels: one two three four five*

> *gl(5,1,15)*

*[1] 1 2 3 4 5 1 2 3 4 5 1 2 3 4 5*

*Levels: 1 2 3 4 5*

The factors thus generated can be combined using the function interaction() to get a resultant combined factor.

> *fac1 <- gl(5,3, labels = c("one", "two", "three", "four", "five"))*

> *fac2 <- gl(5,1,15, labels = c("a", "b", "c", "d", "e", "f", "g", "h", "i", "j",*

                                                    *"k", "l", "m", "n", "o"))*

> *interaction(fac1, fac2)*

*[1] one.a  one.b  one.c  two.d  two.e  two.a  three.b three.c three.d four.e*

*[11] four.a four.b five.c five.d five.e*

*75 Levels: one.a two.a three.a four.a five.a one.b two.b three.b four.b ... five.o*

## 2.7. Strings

Strings are stored in character vectors. Most string manipulation functions act on character vectors. Character vectors can be created using the *c()* function by enclosing the string in double or single quotes. (Generally we follow only double quotes). The *paste()* function can be used to concatenate two strings with  a space in between. If the space need not be shown, we use the function *paste0()*. To have specified separator between the two concatenated string, we use the argument sep in the *paste()* function. The result can be collapsed into one string using the collapse argument.

> *c("String 1", 'String 2')*

*[1] "String 1" "String 2"*

> *paste(c("Pine", "Red"), "Apple")*

*[1] "Pine Apple" "Red Apple"*

*> paste0(c("Pine", "Red"), "Apple")*

*[1] "PineApple" "RedApple"*

*> paste(c("Pine", "Red"), "Apple", sep = "-")*

*[1] "Pine-Apple" "Red-Apple"*

*> paste(c("Pine", "Red"), "Apple", sep = "-", collapse = ", ")*

*[1] "Pine-Apple, Red-Apple"*

The *to String()* function can be used to convert a number vector into a character vector, with the elements separated by a comma and a space. It is possible to specify the width of the print string in this function.

*> x <- c(1:10) ^ 3*

*> x*

*[1]    1    8   27   64  125  216  343  512  729 1000*

*> toString(x)*

*[1] "1, 8, 27, 64, 125, 216, 343, 512, 729, 1000"*

*> toString(x, 18)*

*[1] "1, 8, 27, 64, ...."*

The *cat()* function is also similar to the *paste()* function, but there is little difference in it as shown below.

*> cat(c("Red", "Pine"), "Apple")*

*Red Pine Apple*

The *noquote()* function forces the string outputs not to be displayed with quotes.

*> a <- c("I", "am", "a", "data", "scientist")*

*> a*

*[1] "I"      "am"     "a"      "data"    "scientist"*

*> noquote(a)*

*[1] I       am     a       data     scientist*

The *formatC()* function is used to format the numbers and display them as strings. This function has the arguments *digits, width, format, flag* etc which can be used as below. A slight variation of the function *formatC()* is the function *format()* whose usage is as shown below.

> h <- c(4.567, 8.981, 27.772)

> h

[1]  4.567  8.981 27.772

> formatC(h)

[1] "4.567" "8.981" "27.77"

> formatC(h, digits = 3)

[1] "4.57" "8.98" "27.8"

> formatC(h, digits = 3, width = 5)

[1] " 4.57" " 8.98" " 27.8"

> formatC(h, digits = 3, format = "e")

[1] "4.567e+00" "8.981e+00" "2.777e+01"

> formatC(h, digits = 3, flag = "+")

[1] "+4.57" "+8.98" "+27.8"

> format(h)

[1] " 4.567" " 8.981" "27.772"

> format(h, digits = 3)

[1] " 4.57" " 8.98" "27.77"

> format(h, digits = 3, trim = TRUE)

[1] "4.57" "8.98" "27.77"

The *sprint()* function is also used for formatting strings and passing number values in between the strings. The argument %s in this function stands for a string to be passed. The argument %d and argument %f stands for integer and floating-point number. The usage of this function can be understood by the below example.

> x <- c(1, 2, 3)

> sprintf("The number %d in the list is = %f", x, h)

[1] "The number 1 in the list is = 4.567000"

[2] "The number 2 in the list is = 8.981000"

[3] "The number 3 in the list is = 27.772000"

To print a tab in between text, we can use the *cat()* function with the special character "\t" included in between the text as below. Similarly, if we need to insert a new line in between the text, we use "\n". In this *cat()* function the argument *fill = TRUE* means that after printing the text, the cursor is placed in the next line. Suppose if a back slash has to be used in between the text, it is preceded by another back slash. If we enclose the text in double quotes and if the text contains a double quote in between, it is also preceded by a back slash. Similarly, if we enclose the text in single quotes and if the text contains a single quote in between, it is also preceded by a back slash. If we enclose the text in double quotes and if the text contains a single quote in between, or if we enclose the text in single quotes and if the text contains a double quote in between, it is not a problem (No need for back slash).

> cat("Black\tBerry", fill = TRUE)

Black        Berry

> cat("Black\nBerry", fill = TRUE)

Black

Berry

> cat("Black\\Berry", fill = TRUE)

Black\Berry

> cat("Black\"Berry", fill = TRUE)

Black"Berry

> cat('Black\'Berry', fill = TRUE)

Black'Berry

> cat('Black"Berry', fill = TRUE)

Black"Berry

> *cat("Black'Berry", fill = TRUE)*

*Black'Berry*

The function *toupper()* and *tolower()* are used to convert a string into upper case or lower case respectively. The *substring()* or the *substr()* function is used to cut a part of the string from the given text. Its arguments are the text, starting position and ending position. Both these functions produce the same result.

> *toupper("The cat is on the Wall")*

*[1] "THE CAT IS ON THE WALL"*

> *tolower("The cat is on the Wall")*

*[1] "the cat is on the wall"*

> *substring("The cat is on the wall", 3, 10)*

*[1] "e cat is"*

> *substr("The cat is on the wall", 3, 10)*

*[1] "e cat is"*

> *substr("The cat is on the wall", 5, 10)*

*[1] "cat is"*

The function *strsplit()* does the splitting of a text into many strings based on the splitting character mentioned as argument. In the below example the splitting is done when a space is encountered. It is important to note that this function returns a list and not a character vector as a result.

> *strsplit("I like Bannana, Orange and Pineapple", " ")*

*[[1]]*

*[1] "I"      "like"     "Bannana," "Orange"  "and"      "Pineapple"*

In this same example if the text has to be split when a comma or space is encountered it is mentioned as ",?". This means that the comma is optional and space is mandatory for splitting the given text.

> *strsplit("I like Bannana, Orange and Pineapple", ",? ")*

*[[1]]*

*[1] "I"       "like"      "Bannana"  "Orange"  "and"       "Pineapple"*

The default R's working directory can be obtained using the function *getwd()* and this default directory can be changed using the function *setwd()*. The directory path mentioned in the *setwd()* function should have the forward slash instead of backward slash as in the example below.

> *getwd()*

*[1] "C:/Users/admin/Documents"*

> *setwd("C:/Program Files/R")*

> *getwd()*

*[1] "C:/Program Files/R"*

It is also possible to construct the file paths using the *file.path()* function which automatically inserts the forward slash between the directory names. The function *R.home()* list the home directory where R is installed.

> *file.path("C:", "Program Files", "R", "R-3.3.0")*

*[1] "C:/Program Files/R/R-3.3.0"*

> *R.home()*

*[1] "C:/PROGRA~1/R/R-33~1.0"*

Paths can also be specified by relative terms such as "." denotes current directory, ".." denotes parent directory and "~" denotes home directory. The function *path.expand()* converts relative paths to absolute paths.

> *path.expand(".")*

*[1] "."*

> *path.expand("..")*

*[1] ".."*

> *path.expand("~")*

*[1] "C:/Users/admin/Documents"*

The function *basename()* returns only the file name leaving its directory if specified. On the other hand the function *dirname()* returns only the directory name leaving the file name.

> *filename <- "C:/Program Files/R/R-3.3.0/bin/R.exe"*

> *basename(filename)*

*[1] "R.exe"*

> *dirname(filename)*

*[1] "C:/Program Files/R/R-3.3.0/bin"*

# 2.8. Dates and Times

Dates and Times are common in data analysis and R has a wide range of capabilities for dealing with dates and times.

## 2.8.1. Date and Time Classes

R has three date and time base classes and they are *POSIXct, POSIXlt* and *Date*. *POSIX* is a set of standards that defines how dates and times should be specified and "*ct*" stands for "calendar time". *POSIXlt* stores dates as a list of seconds, minutes, hours, day of month etc. For storing and calculating with dates, we can use *POSIXct* and for extracting parts of dates, we can use *POSXlt*.

The function *Sys.time()* is used to return the current date and time. This returned value is by default in the *POSIXct* form. But, this can be converted to *POSIXlt* form using the function *as.POSIXlt()*. When printed both forms of date and time are displayed in the same manner, but their internal storage mechanism differs. We can also access individual components of a *POSIXlt* date using the dollar symbol or the double brackets as shown below.

> *Sys.time()*

*[1] "2017-05-11 14:31:29 IST"*

> *t <- Sys.time()*

> *t1 <- Sys.time()*

> *t2 <- as.POSIXlt(t1)*

```
> t1
[1] "2017-05-11 14:39:39 IST"
> t2
[1] "2017-05-11 14:39:39 IST"
> class(t1)
[1] "POSIXct" "POSIXt"
> class(t2)
[1] "POSIXlt" "POSIXt"
> t2$sec
[1] 39.20794
> t2[["min"]]
[1] 39
> t2$hour
[1] 14
> t2$mday
[1] 11
> t2$wday
[1] 4
```

The Date class stores the dates as number of days from start of 1970. This class is useful when time is insignificant. The *as.Date()* function can be used to convert a date in other class formats to the Date class format.

```
> t3 <- as.Date(t2)
> t3
[1] "2017-05-11"
```

There are also other add-on packages available in R to handle date and time and they are *date, dates, chron, yearmon, yearqtr, timeDate, ti* and *jul.*

## 2.8.2. Date Conversions

In CSV files the dates will be normally stored as strings and they have to be converted into date and time using any of the packages. For this we need to parse the strings using the function *strptime()* and this returns the date of the format *POSIXlt*. The date format is specified as a string and passed as argument to the *strptime()* function. If the given string does not match the format given in the format string, then it returns NA.

> *date1 <- strptime("22:15:45 22/08/2015", "%H:%M:%S %d/%m/%Y")*

> *date1*

[1] *"2015-08-22 22:15:45 IST"*

> *date2 <- strptime("22:15:45 22/08/2015", "%H:%M:%S %d-%m-%Y")*

> *date2*

[1] NA

In the format string "*%H*" denotes hour in 24 hour system, "*%M*" denotes minutes, "*%S*" denotes second, "*%m*" denotes the number of the month, "*%d*" denotes the day of the month as number, "*%Y*" denotes four digit year.

To convert a date into a string the function *strftime()* is used. This function also takes a date formatting string as argument like *strptime()*. In the format string "*%I*" denotes hour in 12 hours system, "*%p*" denotes AM/PM, "*%A*" denotes the string of day of the week, "*%B*" denotes the string of name of the month.

> *strftime(Sys.Date(),"It's %I:%M%p on %A %d %B, %Y.")*

[1] *"It's 12:00AM on Thursday 11 May, 2017."*

## 2.8.3. Time Zones

It is possible to specify the time zone when parsing a date string using *strptime()* or *strftime()* functions. If this is not specified, the default time zone is taken. The functions *Sys.timezone()* and *Sys.getlocale("LC_TIME")* are used to get the default time zone of the system and the operating system respectively.

> *Sys.timezone()*

*[1] "Asia/Calcutta"*

> *Sys.getlocale("LC_TIME")*

*[1] "English_India.1252"*

Few of the time zones are *UTC* (Universal Time), *IST* (Indian Standard Time), *EST* (Eastern Standard Time), *PST* (Pacific Standard Time), *GMT* (Greenwitch Meridian Time), etc. It is also possible to give manual offset from *UTC* as "*UTC+n*" or "*UTC–n*" to denote west and east parts of *UTC* respectively. Even though it throws warning message, it gives the result correctly.

> *strftime(Sys.time(), tz = "UTC")*

*[1] "2017-05-12 04:59:04"*

> *strftime(Sys.time(), tz = "UTC-5")*

*[1] "2017-05-12 09:59:09"*

*Warning message:*

*In as.POSIXlt.POSIXct(x, tz = tz) : unknown timezone 'UTC-5'*

> *strftime(Sys.time(), tz = "UTC+5")*

*[1] "2017-05-11 23:59:15"*

*Warning message:*

*In as.POSIXlt.POSIXct(x, tz = tz) : unknown timezone 'UTC+5'*

The time zone changes does not happen in *strftime()* function if the date is in *POSIXlt* dates. Hence, it is required to change to *POSIXct* format first and then apply the function.

## 2.8.4. Calculations with Dates and Times

If we add a number to the *POSIXct* or *POSIXlt* classes, it will shift to that many seconds. If we add a number to the *Date* class, it will shift to that many days.

> *ct <- as.POSIXct(Sys.time())*

> *lt <- as.POSIXlt(Sys.time())*

```
> dt <- as.Date(Sys.time())

> ct

[1] "2017-05-12 11:41:54 IST"

> ct + 2500

[1] "2017-05-12 12:23:34 IST"

> lt

[1] "2017-05-12 11:42:15 IST"

> lt + 2500

[1] "2017-05-12 12:23:55 IST"

> dt

[1] "2017-05-12"

> dt + 2

[1] "2017-05-14"
```

Adding two dates, throws error. But subtracting two dates gives the number of days in between the dates. To get the same result, alternatively, the *difftime()* function can be used and in this it is possible to specify the attribute *units* = *"secs"* (or *"mins"* or *"hours"* or *"days"* or *"weeks"*).

```
> dt1 <- as.Date("10/10/1973", "%d/%m/%Y")

> dt1

[1] "1973-10-10"

> dt2 <- as.Date("25/09/2000", "%d/%m/%Y")

> dt2

[1] "2000-09-25"

> diff <- dt2 - dt1

> diff

Time difference of 9847 days

> difftime(dt2, dt1)
```

*Time difference of 9847 days*

*> difftime(dt2, dt1, units = "secs")*

*Time difference of 850780800 secs*

*> difftime(dt2, dt1, units = "mins")*

*Time difference of 14179680 mins*

*> difftime(dt2, dt1, units = "hours")*

*Time difference of 236328 hours*

*> difftime(dt2, dt1, units = "days")*

*Time difference of 9847 days*

*> difftime(dt2, dt1, units = "weeks")*

*Time difference of 1406.714 weeks*

The *seq()* function can be used to generate a sequence of dates. The argument *"by"* can take many options based on the class of the dates specified. We can apply the *mean()* and *summary()* functions on these sequence of dates generate.

*> seq(dt1, dt2, by = "1 year")*

*[1] "1973-10-10" "1974-10-10" "1975-10-10" "1976-10-10" "1977-10-10"*

                                     *"1978-10-10"*

*[7] "1979-10-10" "1980-10-10" "1981-10-10" "1982-10-10" "1983-10-10"*

                                     *"1984-10-10"*

*[13] "1985-10-10" "1986-10-10" "1987-10-10" "1988-10-10" "1989-10-10"*

                                     *"1990-10-10"*

*[19] "1991-10-10" "1992-10-10" "1993-10-10" "1994-10-10" "1995-10-10"*

                                     *"1996-10-10"*

*[25] "1997-10-10" "1998-10-10" "1999-10-10"*

*> seq(dt1, dt2, by = "500 days")*

*[1] "1973-10-10" "1975-02-22" "1976-07-06" "1977-11-18" "1979-04-02"*

*"1980-08-14"*

*[7] "1981-12-27" "1983-05-11" "1984-09-22" "1986-02-04" "1987-06-19"*

*"1988-10-31"*

*[13] "1990-03-15" "1991-07-28" "1992-12-09" "1994-04-23" "1995-09-05"*

*"1997-01-17"*

*[19] "1998-06-01" "1999-10-14"*

*> mean(seq(dt1, dt2, by = "1 year"))*

*[1] "1986-10-10"*

*> summary(seq(dt1, dt2, by = "1 year"))*

| *Min.* | *1st Qu.* | *Median* | *Mean* | *3rd Qu.* | *Max.* |
|---|---|---|---|---|---|
| *"1973-10-10"* | *"1980-04-10"* | *"1986-10-10"* | *"1986-10-10"* | *"1993-04-10"* | *"1999-10-10"* |

The *lubridate* package makes the process of date and time manipulation easier. The *ymd()* function in this package converts any date to the format of year, month and day separated by hyphens.(*Note:* This function requires the date to be specified in the order of year, month and day, but can use any separator as below).

*> install.packages("lubridate")*

*> library(lubridate)*

*> ymd("2000/09/25", "2000-9-25", "2000*9.25")*

*[1] "2000-09-25" "2000-09-25" "2000-09-25"*

If the given date is in other formats that is not in the order of year, month and day, then we have other functions such as *ydm()*, *mdy()*, *myd()*, *dmy()* and *dym()*. These functions can also be accompanied with time by making use of the functions *ymd_h()*, *ymd_hm()* and *ymd_hms()* [similar functions available for *ydm()*, *mdy()*, *myd()*, *dmy()* and *dym()*]. All the parsing functions in the *lubridate* package returns *POSIXct* dates and the default time zone is *UTC*. A function named *stamp()* in the *lubridate* package allows formatting of the dates in a human readable format.

*> dt_format <- stamp("I purchased on Sunday, the 10th of October 2013 at*

*6:00:00 PM")*

*Multiple formats matched: "I purchased on %A, the %dth of %B %Y at %H:%M:%S*

*%Op" (0), "I purchased on %A, the %dth of October %Y at %Om:%H:%M %Op"...*

...

*Using: "I purchased groceries on %A, the %dth of %Om %Y at %H:%M:%S %Op"*

> *dt_to_convert <- strptime("2000-09-25 7:00:00", "%Y-%m-%d %H:%M:%S")*

> *dt_format(dt_to_convert)*

*[1] "I purchased groceries on Monday, the 25th of 09 2000 at 07:00:00 AM"*

The *lubridate* package has three variable types, namely the *"Durations"*, *"Periods"* and *"Intervals"*. The *lubridate* package has the functions, *dyears()*, *dweeks()*, *ddays()*, *dhours()*, *dminutes()*, *dseconds()* etc that specify the duration of year, week, day, hour, minute and second in terms of seconds. The duration of 1 minute is 60 seconds, the duration of 1 hour is 3600 seconds (60 minutes * 60 seconds), the duration of 1 day is 86,400 seconds (24 hours * 60 minutes * 60 seconds), the duration of 1 year is 31,536,000 seconds (365 days * 24 hours * 60 minutes * 60 seconds) and so on. The function *today()* returns the current days date.

> *y <- dyears(1:5)*

> *y*

*[1] "31536000s (~52.14 weeks)" "63072000s (~2 years)"    "94608000s (~3 years)"*

*[4] "126144000s (~4 years)"    "157680000s (~5 years)"*

> *w <- dweeks(1:4)*

> *w*

*[1] "604800s (~1 weeks)" "1209600s (~2 weeks)" "1814400s (~3 weeks)"*

*[4] "2419200s (~4 weeks)"*

> *d <- ddays(1:10)*

> *d*

*[1] "86400s (~1 days)"     "172800s (~2 days)"    "259200s (~3 days)"*

*[4] "345600s (~4 days)"    "432000s (~5 days)"    "518400s (~6 days)"*

*[7] "604800s (~1 weeks)"    "691200s (~1.14 weeks)" "777600s (~1.29 weeks)"*

*[10] "864000s (~1.43 weeks)"*

```
> today() + y
```

*[1] "2018-05-12" "2019-05-12" "2020-05-11" "2021-05-11" "2022-05-11"*

"Periods" specify time spans according to the clock time. The *lubridate* package has the functions, *years(), weeks(), days(), hours(), minutes(), seconds()* etc that specify the period of year, week, day, hour, minute and second in terms of clock time. The exact length of these periods can be realized only if they are added to an instance of date or time.

```
> y <- years(1:7)
> y
```

*[1] "1y 0m 0d 0H 0M 0S" "2y 0m 0d 0H 0M 0S" "3y 0m 0d 0H 0M 0S"*

*"4y 0m 0d 0H 0M 0S"*

*[5] "5y 0m 0d 0H 0M 0S" "6y 0m 0d 0H 0M 0S" "7y 0m 0d 0H 0M 0S"*

```
> today() +y
```

*[1] "2018-05-12" "2019-05-12" "2020-05-12" "2021-05-12" "2022-05-12"*

*"2023-05-12"*

*[7] "2024-05-12"*

"*Intervals*" are defined by the instance of date or time at the beginning and end. They are mostly used for specifying "*Periods*" and "*Durations*" and conversion between "*Periods*" and "*Durations*".

```
> yr <- dyears(5)
> yr
```

*[1] "157680000s (~5 years)"*

```
> as.period(yr)
```

*[1] "5y 0m 0d 0H 0M 0S"*

```
> sdt <- ymd("2017-05-12")
> int <- new_interval(sdt, sdt+yr)
> int
```

*[1] 2017-05-12 UTC--2022-05-11 UTC*

The operator "*%--%*" is used for defining intervals and the operator "*%within%*" is used for checking if a given date is within the given interval.

> *intv <- ymd("1973-10-10") %--% ymd("2000-09-25")*

> *intv*

*[1] 1973-10-10 UTC--2000-09-25 UTC*

> *ymd("1979-12-12") %within% intv*

*[1] TRUE*

The function *with_tz()* can be used to change the time zone of a date (correctly handles *POSIXlt* dates) and the function *force_tz()* is used for updating incorrect time zones.

> *with_tz(Sys.time(), tz = "America/Los_Angeles")*

*[1] "2017-05-12 06:44:14 PDT"*

> *with_tz(Sys.time(), tz = "Asia/Kolkata")*

*[1] "2017-05-12 19:14:29 IST"*

The functions *floor_date()* and *ceiling_date()* can be used to find the lower and upper limit of a given date as below.

> *floor_date(today(), "year")*

*[1] "2017-01-01"*

> *ceiling_date(today(), "year")*

*[1] "2018-01-01"*

> *floor_date(today(), "month")*

*[1] "2017-05-01"*

> *ceiling_date(today(), "month")*

*[1] "2017-06-01"*

## ❖   HIGHLIGHTS

- The basic data types in R are Numeric, Integer, Complex, Logical and Character.

- To create an integer variable and force a numeric value into an integer in R, the *as.integer()* function is invoked.
- A *paste()* function can be used to concatenate two character values.
- The *substr()* function can be used to extract a substring from a given string.
- The *sub()* function is used to replace the first occurrence of a string with another string.
- The *vector()* function creates a vector of a specified type and length.
- The *seq()* function allows to generate sequences.
- The function *length()* can be used to find the length of the vector.
- The *which()* function returns the elements of the vector which satisfies the condition specified within this function.
- To create an array the *array()* function can be used.
- A matrix can also be created using the *array()* function where the dimension of the array is two.
- The columns of two matrices can be combined using the *cbind()* function.
- The rows of two matrices can be combined using the *rbind()* function.
- Lists can be created using the *list()* function.
- To get the preview or the first few records of a data frame along with the header the *head()* function can be used.
- The *merge()* function can be applied to merge two data frames provided they have common column names.
- Factors are used to store categorical data and they behave like strings sometimes and integers sometimes.
- At the atomic level a factor can be created using the *factor()* function.
- The function *gl()* can be used to create a factor.
- The factors can be combined using the function *interaction()*.
- The default R's working directory can be obtained using the function *getwd()*.
- The default directory can be changed using the function *setwd()*.
- The function *R.home()* list the home directory where R is installed.

- R has three date and time base classes and they are POSIXct, POSIXlt and Date.
- The function *Sys.time()* is used to return the current date and time.
- The *as.Date()* function can be used to convert a date in other class formats to the Date class format.
- The functions *Sys.timezone()* is used to get the default time zone of the system.
- The *lubridate* package has the functions, *dyears()*, *dweeks()*, *ddays()*, *dhours()*, *dminutes()*, *dseconds()* etc that specify the duration of year, week, day, hour, minute and second in terms of seconds.
- The *lubridate* package has the functions, *years()*, *weeks()*, *days()*, *hours()*, *minutes()*, *seconds()* etc that specify the period of year, week, day, hour, minute and second in terms of clock time.

# CHAPTER 3

# DATA PREPARATION

❖ **OBJECTIVES**

On completion of this Chapter you will be able to:

- know about the default datasets available in R
- know how to import and export CSV files in R
- know how to import unstructured data files into R
- know how to import XML and HTML files into R
- know how to import JASON and YAML files into R
- know how to import and export excel files in R
- know how to import SAS, SPSS and MATLAB files into R
- know how to import web data files into R
- understand the concept of accessing various databases from R
- manipulate string data
- manipulate data frames
- understand how to melt and cast data in data frames
- understand how the grouping functions are applied on the data in R

## 3.1. Datasets

R has many datasets built in. R can read data from variety of other data sources and in variety of formats. One of the packages in R is *datasets* which is filled with example datasets. Many other packages also contain datasets. We can see all the datasets available in the loaded packages using the *data()* function.

To access a particular dataset use the *data()* function with its argument as the dataset name enclosed within double quotes and the second optional argument being the package name in which the dataset is present (This second argument is required only if the particular package is not loaded). The invoked dataset can be listed just like a data frame using the *head()* function.

> *data("kidney", package = "survival")*

> *head(kidney)*

|   | id | time | status | age | sex | disease | frail |
|---|----|------|--------|-----|-----|---------|-------|
| 1 | 1 | 8 | 1 | 28 | 1 | Other | 2.3 |
| 2 | 1 | 16 | 1 | 28 | 1 | Other | 2.3 |
| 3 | 2 | 23 | 1 | 48 | 2 | GN | 1.9 |

....

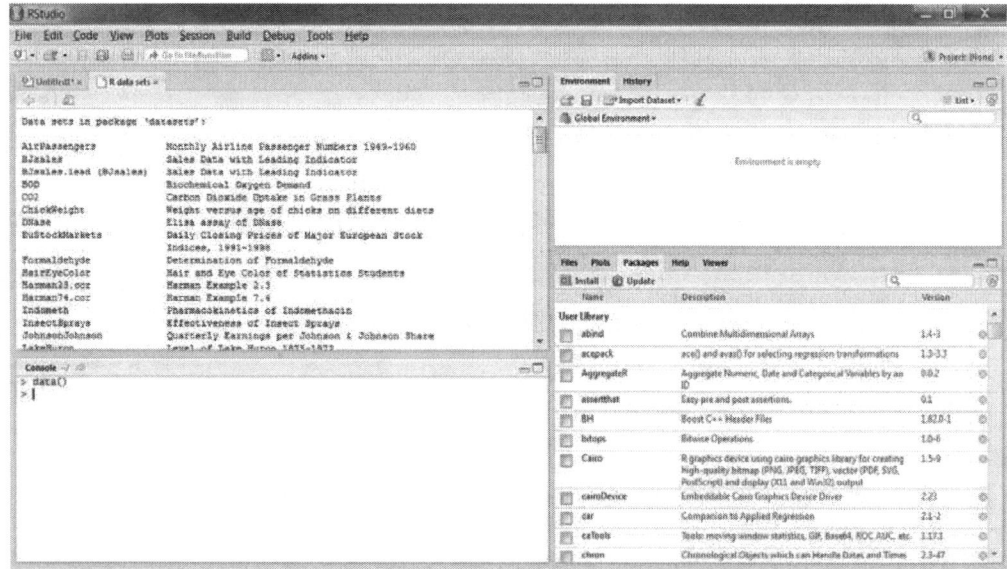

**Figure 3.1**    R-Studio Showing the List of Datasets

## 3.2. Importing and Exporting Files

### 3.2.1. Text and CSV Files

Text documents have several formats. Common format are CSV (Comma Separated Values), XML (Extended Markup Language), JSON (JavaScript Object Notation) and YAML. An example of an unstructured text data is a book.

Comma Separated Values (CSV) Files is a spreadsheet like data stored with comma delimited values. The *read.table()* function reads these files and stores the result in a data frame. If the data has header, it is required to pass the argument *header = TRUE* to the *read.table()* function. The argument *fill = TRUE* makes the *read.table()* function substitute NA values for the missing fields. The *system.file()* function is used to locate files that are inside a package. In the below example *"extdata"* is the folder name and the package name is *"learning"* and the file name is *"RedDeerEndocranialVolume,dlm"* The *str()* function takes the data frame name as the argument and lists the structure of the dataset stored in the data frame.

```
> install.packages("learningr")
> library(learningr)
> deer_file <- system.file("extdata","RedDeerEndocranialVolume.dlm",
                                                    package = "learningr")
> deer_data <- read.table(deer_file, header=TRUE, fill=TRUE)
> str(deer_data)
'data.frame':        33 obs. of  8 variables:
$ SkullID    : Factor w/ 33 levels "A4","B11","B12",..: 14 2 17 16 15 13 10 11
                                                                 19 3 ...
$ VolCT      : int  389 389 352 388 375 325 346 302 379 410 ...
$ VolBead    : int  375 370 345 370 355 320 335 295 360 400 ...
$ VolLWH     : int  1484 1722 1495 1683 1458 1363 1250 1011 1621 1740 ...
$ VolFinarelli: int  337 377 328 377 328 291 289 250 347 387 ...
$ VolCT2     : int  NA NA NA NA NA NA 346 303 375 413 ...
```

$ VolBead2   : int  NA NA NA NA NA NA 330 295 365 395 ...

$ VolLWH2   : int  NA NA NA NA NA NA 1264 1009 1647 1728 ...

The column names and row names are listed by default and if the row names are not given in the dataset, the rows are simply numbered 1, 2, 3 and so on. The arguments specify how the file will be read. The argument *sep* determines the character to use as separator between fields. The *nrow* argument specifies the lines of data to read. The argument *skip* specifies the number of lines to skip at the start of the file. For the functions *read.table()* and *read.csv()* the default separator is set to comma and they assume the data has header row. The function *read.csv2()* uses the semicolon as the separator and comma instead of decimals. The *read.delim()* function imports the tab-delimited files with full stops for decimal places. The *read.delim2()* function imports the tab-delimited files with commas for decimal places.

```
> read.csv(deer_file, header=FALSE, skip = 3, nrow = 2)
                       V1
1 DIC90  352     345     1495    328
2 DIC83  388     370     1683    377
> head(deer_data)
    SkullID VolCT VolBead VolLWH VolFinarelli VolCT2 VolBead2 VolLWH2
1   DIC44  389   375   1484       337   NA    NA    NA
2   B11   389   370   1722       377   NA    NA    NA
3   DIC90  352   345   1495       328   NA    NA    NA
    ....
```

The *colbycol* and *sqldf* packages contain functions that allow to read part of the CSV file into R. These are useful when we don't need all the columns or all the rows. For low-level control we can use the *scan()* function to import CSV file. For data exported from other languages we may need to pass the *na.strings* argument to the *read.table()* function to replace the missing values. If the data is exported from SQL, we use *na.strings* = "NULL" and if the data is exported from SAS or Stata, we use *na.strings* = ".". If the data is exported from Excel we use the *na.strings* = *c("","#N/A", "#DIV/0!", "#NUM!")*.

Writing data from R into a file is easier than reading files into R. For this we use the functions *write.table()* and *write.csv()*. These functions take a data frame and a file path as arguments. They also have arguments to specify if we need not include row names in the output file or to specify the character encoding of the output file.

> *write.csv(deer_data,"F:/deer.csv", row.names = FALSE, fileEncoding = "utf8")*

## 3.2.2. Unstructured Files

If the file structure is week, it is easier to read the file as lines of text using the function *readLines()* and then parse the contents. The *readLines()* function accepts a path to the file as the argument. Similarly, the *writeLines()* function takes a text line or a character vector and the file name as argument and writes the text to the file.

> *tempest <- readLines("F:/Tempest.txt")*

> *tempest*

[1] "The writing of Prefaces to Plays was probably invented by some very"

[2] "ambitious Poet, who never thought he had done enough: Perhaps by ome"

[3] "Ape of the French Eloquence, which uses to make a business of a Letter of"

....

> *writeLines("This book is about a story by Shakespeare", "F:/story.csv")*

## 3.2.3. XML and HTML Files

XML files are used for storing nested data. Few of them are RSS (Really Simple Syndication) feeds, SOAP (Simple Object Access Protocols) and XHTML Web Pages. To read the XML files, the *XML* package has to be installed. When an XML file is imported, the result can be stored using the internal nodes or the R nodes. If the result is stored using internal nodes, it allows to query the node tree using the *XPath* language (used for interrogating XML documents). The XML file can be imported using the function *xmlParse()* function. This function can take the argument *useInternalNodes = FALSE* to use R-level nodes instead of the internal nodes while importing the XML files. But, this is set by default by the *xml TreeParse()* function.

```
> install.packages("XML")
> library(XML)

> xml_file <- system.file("extdata", "options.xml", package = "learningr")
> r_options <- xmlParse(xml_file)

> xmlParse(xml_file, useInternalNodes = FALSE)
> xmlTreeParse(xml_file)
```

The functions for importing HTML pages are *htmlParse()* and *htmlTreeParse()* and they behave same as the *xmlParse()* and *xmlTreeParse()* functions.

## 3.2.4. JASON and YAML Files

The two packages dealing with JSON data are *RJSONIO* and *rjson* and the best of these is the *RJSONIO*. The function used to import the JSON file is *fromJSON()* and the function used to export the JSON file is *toJSON()*. The *yaml* package has two functions for importing YAML data and they are *yaml.load()* and *yaml.load_file()*. The function *as.yaml()* performs the task of converting R objects to YAML strings.

Many softwares store their data in binary formats which are smaller in size than the text files. They hence provide performance gains at the expense of human readability.

## 3.2.5. Excel Files

Excel is the world's most powerful data analysis tool and its document formats are XLX and XLSX. Spreadsheets can be imported with the functions *read.xlsx()* and *read.xlsx2()*. The *colClasses* argument determines what class each column should have in the resulting data frame and this argument is optional in the above functions. To write to an excel file from R we use the function *write.xlsx2()* that takes the data frame and the file name as arguments. There is another package *xlsReadWrite* that does the same function of the *xlsx* package but this one works only in 32-bit R installations and only on windows.

```
> install.packages("xlsx")
> library(xlsx)
> logfile <- read.xlsx2("F:/Log2015.xls", sheetIndex = 1, startRow = 2, endrow = 72,
         colIndex = 1:5, colClasses = c("character", "numeric", "character",
                                        "character", "integer"))
```

## 3.2.6. SAS, SPSS and MATLAB Files

The files from a statistical package are imported using the foreign package. The *read.ssd()* function is used to read SAS datasets and the *read.dta()* function is used to read *Stata DTA* files. The *read.spss()* function is used to import the *SPSS data files*. Similarly, these files can be written with the *write.foreign()* function. The *MATLAB binary data files* can be read and written using the *readMat()* and *writeMat()* functions in the *R.matlab* package. The files in picture formats can be read via the *jpeg, png, tiff, rtiff* and *readbitmap* packages.

## 3.2.7. Web Data

R has ways to import data from web sources using Application Programming Interface (API). For example the World Bank makes its data available using the WDI package and the Polish government data can be accessed using the SmarterPoland package. The twitter package provides access to Twitter's users and their tweet.

The *read.table()* function can accept URL rather than a local file. Accessing a large file from internet can be slow and if the file is required frequently, it is better to download the file using the *download.file()* function and create a local copy and then import that.

```
> cancer_url <- "http://repository.seasr.org/Datasets/UCI/csv/breast-cancer.csv"
> cancer_data <- read.csv(cancer_url)
> str(cancer_data)
'data.frame': 287 obs. of 10 variables:
$ age      : Factor w/ 7 levels "20-29","30-39",..: 7 3 4 4 3 3 4 4 3 3 ...
$ menopause : Factor w/ 4 levels "ge40","lt40",..: 4 3 1 1 3 3 3 1 3 3 ...
```

*$ tumor.size : Factor w/ 12 levels "0-4","10-14",..: 12 3 3 7 7 6 5 8 2 1 ...*

*$ inv.nodes : Factor w/ 8 levels "0-2","12-14",..: 8 1 1 1 1 5 5 1 1 1 ...*

*$ node.caps : Factor w/ 4 levels "","no","String",..: 3 4 2 2 4 4 2 2 2 2 ...*

*$ deg.malig : Factor w/ 4 levels "1","2","3","String": 4 3 1 2 3 2 2 3 2 2 ...*

*$ breast : Factor w/ 3 levels "left","right",..: 3 2 2 1 2 1 2 1 1 2 ...*

*$ breast.quad: Factor w/ 7 levels "","central","left_low",..: 7 4 2 3 3 6 4 4 4 5*

*$ irradiat : Factor w/ 3 levels "no","String",..: 2 1 1 1 3 1 3 1 1 1 ...*

*$ Class : Factor w/ 3 levels "no-recurrence-events",..: 3 2 1 2 1 2 1 1 1 1*

*> local_copy <- "cancer.csv"*

*> download.file(cancer_url, local_copy)*

> *trying URL 'http://repository.seasr.org/Datasets/UCI/csv/breast-cancer.csv'*

> *Content type 'application/octet-stream' length 18804 bytes (18 KB)*

> *downloaded 18 KB*

*> cancer_data <- read.csv(local_copy)*

# 3.3. Accessing Databases

R can connect to all database management systems (DBMS) like *SQLite*, *MySQL*, *MariaDB*, *PostgreSQL* and *Oracle* using the *DBI* package. We need to install and load the *DBI* package and the backend package *RSQLite*. Define a database driver of type *SQLite* using the function *dbDriver()* and setup a connection to the database using the function *dbConnect()*. To retrieve data from the databases you write a query as a string containing SQL commands and send it to the database with the function *dbGetQuery()*.

*> install.packages("DBI")*

*> install.packages("RSQLite")*

*> library(DBI)*

*> library(RSQLite)*

*> driver <- dbDriver("SQLite")*

---

> *db_file <- system.file("extdata", "crabtag.sqlite", package = "learningr")*

> *conn <- dbConnect(driver, db_file)*

> *query <- "SELECT * FROM IdBlock"*

> *id_block <- dbGetQuery(conn, query)*

> *id_block*

    *Tag ID Firmware Version No Firmware Build Level*

    1     A03401          2         70

Alternatively, the function *dbReadTable()* reads a table from the connected database and the function *dbListTables()* can list all the tables in the database.

> *dbReadTable(conn, "idblock")*

    *Tag.ID Firmware.Version.No Firmware.Build.Level*

    1     A03401          2         70

> *dbListTables(conn)*

    *[1] "Daylog"       "DeploymentNotebook" "IdBlock"*

    *[4] "LifetimeNotebook"  "TagNotebook"*

The function *dbDisconnect()* is used for disconnecting and unloading the driver and the function *dbUnloadDriver()* is used to unload the defined database driver.

> *dbDisconnect(conn)*

> *dbUnloadDriver(driver)*

For MySQL database we need to load the *RMySQL* package and set the driver type to be *"MySQL"*. The PostgreSQL, Oracle and JDBC databases need the *PostgreSQL*, *ROracle* and *RJDBC* packages respectively. To connect to an SQL Server or Access databases, the *RODBC* package needs to be loaded. In this package, the function *odbcConnect()* is used to connect to the database and the function *sqlQuery()* is used to run a query and the function *odbcClose()* is used to close and cleanup the database connections. There are not much matured methods to access the NoSQL (Not only SQL) databases (lightweight databases – scalable than traditional SQL relational databases). To access the *MongoDB* database the packages *RMongo* and *rmongodb* are used. The database *Cassandra* can be accessed using the package *RCassandra*.

# 3.4. Data Cleaning and Transforming

## 3.4.1. Manipulating Stings

In some datasets or data frames logical values are represented as "Y" and "N" instead of *TRUE* and *FALSE*. In such cases it is possible to replace the string with correct logical value as in the example below.

```
> a <- c(1,2,3)
> b <- c("A", "B", "C")
> d <- c("Y", "N", "Y")
> df1 <- data.frame(a, b, d)
> df1
```

|   | a | b | d |
|---|---|---|---|
| 1 | 1 | A | Y |
| 2 | 2 | B | N |
| 3 | 3 | C | Y |

```
convt <- function(x)
{
y <- rep.int(NA, length(x))
y[x == "Y"] <- TRUE
y[x == "N"] <- FALSE
y
}
> df1$d <- convt(df1$d)
> df1
```

|   | a | b | d |
|---|---|---|---|
| 1 | 1 | A | TRUE |
| 2 | 2 | B | FALSE |
| 3 | 3 | C | TRUE |

The functions *grep()* and *grepl()* are used to find a pattern in a given text and the functions *sub()* and *gsub()* are used to replace a pattern with another in a given text. The above four functions belong to the *base* package, but the package *stringr* consists of many such string manipulation functions. The function *str_ detect()* in the *stringr* package does the same function of detecting the presence of a given pattern in the given text. We can also use the function *fixed()* to mention if the string that we are searching for is a fixed one.

> *grep("my", "This is my pen")*

*[1] 1*

> *grepl("my", "This is my pen")*

*[1] TRUE*

> *sub("my", "your","This is my pen")*

*[1] "This is your pen"*

> *gsub("my", "your","This is my pen")*

*[1] "This is your pen"*

> *str_detect("This is my pen", "my")*

*[1] TRUE*

> *str_detect("This is my pen", fixed("my"))*

*[1] TRUE*

In the function *str_detect()*, it is possible to specify the search pattern with a pipe symbol " | " to denote, that we need to find either of the two patterns specified. That is we may be looking for the presence of "," or "*and*" in the given text as shown below.

> *str_detect("I like mangoes, oranges and pineapples", ", | and")*

*[1] TRUE*

The function *str_split()* is used to split a given text based on the pattern specified as below. This function returns a vector. But the function *str_split_fixed()* can be used to split the given text into fixed number of strings based on the specified patterns. This function returns a matrix.

> *str_split("I like mangoes, oranges and pineapples", ",|and")*

*[[1]]*

*[1] "I like mangoes" " oranges "      " pineapples"*

> *str_split_fixed("I like mangoes, oranges and pineapples", ",|and", n = 3)*

    *[,1]*                 *[,2]*         *[,3]*

*[1,] "I like mangoes" " oranges " " pineapples"*

The function *str_count()* can be used to count the number of occurrence of a given pattern in the given text.

> *str_count("I like mangoes, oranges and pineapples", "a|o")*

*[1] 6*

> *str_count("I like mangoes, oranges and pineapples", "s")*

*[1] 3*

The function *str_replace()* can be used to replace the specified pattern with another pattern in the given text. This function will only replace the first occurrence of the pattern. Hence, to replace all the occurrences of the pattern we use the function *str_replace_all()*. In these functions, to denote multiple patterns to be replaced, they can be placed within square brackets. This means it should replace all that matches these characters specified within the square brackets.

> *str_replace("I like mangoes, oranges and pineapples", "s", "sss")*

*[1] "I like mangoesss, oranges and pineapples"*

> *str_replace_all("I like mangoes, oranges and pineapples", "s", "sss")*

*[1] "I like mangoesss, orangesss and pineapplesss"*

> *str_replace_all("I like mangoes, oranges and pineapples", "[ao]", "-")*

*[1] "I like m-ng-es, -r-nges -nd pine-pples"*

In the example below, the various ways of storing the gender values are transformed into one way, ignoring the case differences. This is done using the *str_replace()* function and the *fixed()* functions that ignores the case.

> *gender <- c("MALE", "Male", "male", "FEMALE", "Female", "female")*

> *clean_gender <- str_replace(gender, fixed("male", ignore_case = TRUE), "Male")*

> *clean_gender <- str_replace(clean_gender, fixed("female", ignore_case = TRUE),*

*male")*

> *clean_gender*

*[1] "Male"   "Male"   "Male"   "Female" "Female" "Female"*

## 3.4.2. Manipulating Data Frames

To add a column to a data frame, we can use the below command to achieve this.

> *name <- c("Jhon", "Peter", "Mark")*

> *start_date <- c("1980-10-10", "1999-12-12", "1990-04-05")*

> *end_date <- c("1989-03-08", "2004-09-20", "2000-09-25")*

> *service <- data.frame(name, start_date, end_date)*

> *service*

|   | name | start_date | end_date |
|---|------|------------|----------|
| 1 | Jhon | 1980-10-10 | 1989-03-08 |
| 2 | Peter | 1999-12-12 | 2004-09-20 |
| 3 | Mark | 1990-04-05 | 2000-09-25 |

> *service$period <- as.Date(service$end_date) - as.Date(service$start_date)*

> *service*

|   | name | start_date | end_date | period |
|---|------|------------|----------|--------|
| 1 | Jhon | 1980-10-10 | 1989-03-08 | 3071 days |
| 2 | Peter | 1999-12-12 | 2004-09-20 | 1744 days |
| 3 | Mark | 1990-04-05 | 2000-09-25 | 3826 days |

The same can be achieved using the function *with()* as below.

> *service$period <- with(service, as.Date(end_date) - as.Date(start_date))*

> *service*

| | name | start_date | end_date | period |
|---|---|---|---|---|
| 1 | Jhon | 1980-10-10 | 1989-03-08 | 3071 days |
| 2 | Peter | 1999-12-12 | 2004-09-20 | 1744 days |
| 3 | Mark | 1990-04-05 | 2000-09-25 | 3826 days |

Another way of doing the same is using the function *within()*. But, the difference lies when there are multiple columns to be added to a data frame, we can easily do the same using the *within()* function in a single command and this is not possible using the *with()* function.

```
> service <- within(service,
  {
    period <- as.Date(end_date) - as.Date(start_date)
    highperiod <- period > 2000
  })
> service
```

| | name | start_date | end_date | period | highperiod |
|---|---|---|---|---|---|
| 1 | Jhon | 1980-10-10 | 1989-03-08 | 3071 days | TRUE |
| 2 | Peter | 1999-12-12 | 2004-09-20 | 1744 days | FALSE |
| 3 | Mark | 1990-04-05 | 2000-09-25 | 3826 days | TRUE |

The *mutate()* function in the *plyr* package also does the same function as the function *within()*, but the syntax is slightly different.

```
> library(plyr)
> service <- mutate(service,
  {
    period = as.Date(end_date) - as.Date(start_date)
    highperiod = period > 2000
  })
> service
```

|   | name | start_date | end_date | period | highperiod |
|---|------|-----------|----------|--------|-----------|
| 1 | Jhon | 1980-10-10 | 1989-03-08 | 3071 days | TRUE |
| 2 | Peter | 1999-12-12 | 2004-09-20 | 1744 days | FALSE |
| 3 | Mark | 1990-04-05 | 2000-09-25 | 3826 days | TRUE |

The function *complete.cases()* returns the number of rows in a data frame that is free of missing values. The function *na.omit()* will remove the rows with missing values in a data frame. And the function *na.fail()* throws an error message if the data frame contains any missing values.

```
> crime.data <- read.csv("F:/Crimes.csv")
> nrow(crime.data)
[1] 65535
> complete <- complete.cases(crime.data)
> nrow(crime.data[complete, ])
[1] 63799
> clean.crime.data <- na.omit(crime.data)
> nrow(clean.crime.data)
[1] 63799
```

A data frame can be transformed by choosing few of the columns and ignoring the remaining, but considering all the rows as in the example below.

```
> crime.data <- read.csv("F:/Crimes.csv")
> colnames(crime.data)
[1] "CASE."              "DATE..OF.OCCURRENCE"  "BLOCK"
[4] "IUCR"               "PRIMARY.DESCRIPTION"
"SECONDARY.DESCRIPTION"
[7] "LOCATION.DESCRIPTION" "ARREST"              "DOMESTIC"
[10] "BEAT"               "WARD"                 "FBI.CD"
[13] "X.COORDINATE"       "Y.COORDINATE"         "LATITUDE"
[16] "LONGITUDE"          "LOCATION"
```

```
> crime.data1 <- crime.data[, 1:6]
> colnames(crime.data1)
```
[1] "CASE."          "DATE..OF.OCCURRENCE"  "BLOCK"
[4] "IUCR"           "PRIMARY.DESCRIPTION"
"SECONDARY.DESCRIPTION"

Alternatively, the data frame can be transformed by selecting only the required rows and retaining all columns of a data frame as in the example below.

```
> nrow(crime.data)
```
[1] 65535
```
> crime.data2 <- crime.data[1:10,]
> nrow(crime.data2)
```
[1] 10

The function *sort()* sorts the given vector of numbers or strings. It generally sorts from smallest to largest, but this can be altered using the argument *decreasing = TRUE*.

```
> x <- c(5, 10, 3, 15, 6, 8)
> sort(x)
```
[1]  3  5  6  8 10 15
```
> sort(x, decreasing = TRUE)
```
[1] 15 10  8  6  5  3
```
> y <- c("X", "AB", "Deer", "For", "Moon")
> sort(y)
```
[1] "AB"   "Deer" "For"  "Moon" "X"
```
> sort(y, decreasing = TRUE)
```
[1] "X"    "Moon" "For"  "Deer" "AB"

The function *order()* is the inverse of the *sort()* function. It returns the index of the vector elements in the order as below. But, *x[order(x)]* is same as *sort(x)*. This can be seen by the use of the *identical()* function.

> *order(x)*

*[1] 3 1 5 6 2 4*

> *x[order(x)]*

*[1]  3  5  6  8 10 15*

> *identical(sort(x), x[order(x)])*

*[1] TRUE*

The *order()* function is more useful than the *sort()* function as it can be used to manipulate the data frames easily.

> *name <- c("Jhon", "Peter", "Mark")*

> *start_date <- c("1980-10-10", "1999-12-12", "1990-04-05")*

> *end_date <- c("1989-03-08", "2004-09-20", "2000-09-25")*

> *service <- data.frame(name, start_date, end_date)*

> *service*

|   | *name* | *start_date* | *end_date* |
|---|--------|--------------|------------|
| *1* | *Jhon* | *1980-10-10* | *1989-03-08* |
| *2* | *Peter* | *1999-12-12* | *2004-09-20* |
| *3* | *Mark* | *1990-04-05* | *2000-09-25* |

> *startdt <- order(service$start_date)*

> *service.ordered <- service[startdt, ]*

> *service.ordered*

|   | *name* | *start_date* | *end_date* |
|---|--------|--------------|------------|
| *1* | *Jhon* | *1980-10-10* | *1989-03-08* |
| *3* | *Mark* | *1990-04-05* | *2000-09-25* |
| *2* | *Peter* | *1999-12-12* | *2004-09-20* |

The *arrange()* function of the *plyr* package does the same function as above.

> *library(plyr)*

> *arrange(service, start_date)*

|   | name | start_date | end_date |
|---|------|-----------|----------|
| 1 | Jhon | 1980-10-10 | 1989-03-08 |
| 2 | Mark | 1990-04-05 | 2000-09-25 |
| 3 | Peter | 1999-12-12 | 2004-09-20 |

The *rank()* function lists the rank of the elements in a vector or a data frame. By specifying the argument *ties.method* = *"first"*, a rank need not be shared among more than one element with the same value.

```
> x <- c(9, 5, 4, 6, 4, 5)
> rank(x)
[1] 6.0 3.5 1.5 5.0 1.5 3.5
> rank(x, ties.method = "first")
[1] 6 3 1 5 2 4
```

The SQL statements can be executed from R and the results can be obtained as in any other database. The package *sqldf* needs to be installed to manipulate the data frames or datasets using SQL.

```
> install.packages("sqldf")
> library(sqldf)
> query <- "SELECT * FROM iris WHERE Species = 'setosa'"
> sqldf(query)
```

## 3.4.3. Data Reshaping

Data Reshaping in R is about changing the way data is organized into rows and columns. Most of the time data processing in R is done by taking the input data as a data frame. It is easy to extract data from the rows and columns of a data frame. But there are situations when we need the data frame in a different format than what we received. R has few functions to split, merge and change the columns to rows and vice-versa in a data frame.

The *cbind()* function can be used to join multiple vectors to create a data frame. We can also merge two data frames using the *rbind()* function.

```
> city <- c("Tampa","Seattle","Hartford","Denver")

> state <- c("FL","WA","CT","CO")

> zipcode <- c(33602,98104,06161,80294)

> addresses <- cbind(city,state,zipcode)

> addresses
```

|       | city        | state  | zipcode   |
|-------|-------------|--------|-----------|
| [1,]  | "Tampa"     | "FL"   | "33602"   |
| [2,]  | "Seattle"   | "WA"   | "98104"   |
| [3,]  | "Hartford"  | "CT"   | "6161"    |
| [4,]  | "Denver"    | "CO"   | "80294"   |

```
> new.address <- data.frame(

+    city = c("Lowry","Charlotte"),

+    state = c("CO","FL"),

+    zipcode = c("80230","33949"),

+    stringsAsFactors = FALSE

+ )

> print(new.address)
```

|   | city      | state | zipcode |
|---|-----------|-------|---------|
| 1 | Lowry     | CO    | 80230   |
| 2 | Charlotte | FL    | 33949   |

```
> all.addresses <- rbind(addresses,new.address)

> all.addresses
```

|   | city     | state | zipcode |
|---|----------|-------|---------|
| 1 | Tampa    | FL    | 33602   |
| 2 | Seattle  | WA    | 98104   |
| 3 | Hartford | CT    | 6161    |
| 4 | Denver   | CO    | 80294   |

| 5 | Lowry | CO | 80230 |
| 6 | Charlotte | FL | 33949 |

The *merge()* function can be used to merge two data frames. The merging requires the data frames to have same column names on which the merging is done. In the example below, we consider the data sets about *Diabetes in Pima Indian Women* available in the library named "MASS". The two datasets are merged based on the values of *blood pressure ("bp")* and *body mass index ("bmi")*. On choosing these two columns for merging, the records where values of these two variables match in both data sets are combined together to form a single data frame.

```
> library(MASS)
> head(Pima.te)
```

|  | npreg | glu | bp | skin | bmi | ped | age | type |
|---|---|---|---|---|---|---|---|---|
| 1 | 6 | 148 | 72 | 35 | 33.6 | 0.627 | 50 | Yes |
| 2 | 1 | 85 | 66 | 29 | 26.6 | 0.351 | 31 | No |
| 3 | 1 | 89 | 66 | 23 | 28.1 | 0.167 | 21 | No |

...

```
> head(Pima.tr)
```

|  | npreg | glu | bp | skin | bmi | ped | age | type |
|---|---|---|---|---|---|---|---|---|
| 1 | 5 | 86 | 68 | 28 | 30.2 | 0.364 | 24 | No |
| 2 | 7 | 195 | 70 | 33 | 25.1 | 0.163 | 55 | Yes |
| 3 | 5 | 77 | 82 | 41 | 35.8 | 0.156 | 35 | No |

...

```
> nrow(Pima.te)
[1] 332
> nrow(Pima.tr)
[1] 200
> merged.Pima <- merge(x = Pima.te, y = Pima.tr,
+               by.x = c("bp", "bmi"),
+               by.y = c("bp", "bmi")
```

+ )

> head(merged.Pima)

| | bp | bmi | npreg.x | glu.x | skin.x | ped.x | age.x | type.x | npreg.y | glu.y | skin.y |
|---|---|---|---|---|---|---|---|---|---|---|---|
| 1 | 60 | 33.8 | 1 | 117 | 23 | 0.466 | 27 | No | 2 | 125 | 20 |
| 2 | 64 | 29.7 | 2 | 75 | 24 | 0.370 | 33 | No | 2 | 100 | 23 |
| 3 | 64 | 31.2 | 5 | 189 | 33 | 0.583 | 29 | Yes | 3 | 158 | 13 |

...

| | ped.y | age.y | type.y |
|---|---|---|---|
| 1 | 0.088 | 31 | No |
| 2 | 0.368 | 21 | No |
| 3 | 0.295 | 24 | No |

...

> nrow(merged.Pima)

[1] 17

One of the most interesting aspects of R programming is about changing the shape of the data in multiple steps to get a desired shape. The functions used to do this are called *melt()* and *cast()*. We consider the dataset called *ships* present in the library called "MASS".

> library(MASS)

> head(ships)

| | type | year | period | service | incidents |
|---|---|---|---|---|---|
| 1 | A | 60 | 60 | 127 | 0 |
| 2 | A | 60 | 75 | 63 | 0 |
| 3 | A | 65 | 60 | 1095 | 3 |

...

Now we melt the data using the *melt()* function in the package *reshape2* to organize it, converting all columns other than type and year into multiple rows.

```
> library(reshape2)
> molten.ships <- melt(ships, id = c("type","year"))
> head(molten.ships)
```

|   | type | year | variable | value |
|---|------|------|----------|-------|
| 1 | A | 60 | period | 60 |
| 2 | A | 60 | period | 75 |
| 3 | A | 65 | period | 60 |
| 4 | A | 65 | period | 75 |
| 5 | A | 70 | period | 60 |
| 6 | A | 70 | period | 75 |

```
> nrow(molten.ships)
[1] 120
> nrow(ships)
[1] 40
```

We can cast the molten data into a new form where the aggregate of each type of ship for each year is created. It is done using the *cast()* function.

```
> recasted.ship <- cast(molten.ships, type+year~variable,sum)
> head(recasted.ship)
```

|   | type | year | period | service | incidents |
|---|------|------|--------|---------|-----------|
| 1 | A | 60 | 135 | 190 | 0 |
| 2 | A | 65 | 135 | 2190 | 7 |
| 3 | A | 70 | 135 | 4865 | 24 |
| 4 | A | 75 | 135 | 2244 | 11 |
| 5 | B | 60 | 135 | 62058 | 68 |
| 6 | B | 65 | 135 | 48979 | 111 |

## 3.4.4. Grouping Functions

R has many apply functions such as *apply()*, *lapply()*, *sapply()*, *vapply()*, *mapply()*, *rapply()*, *tapply()*, *aggregate()* and *by()*. Function *lapply()* is a list apply which acts on a list or vector and returns a list. Function *sapply()* is a simple *lapply()* function defaults to returning a vector or matrix when possible. Function *vapply()* is a verified *apply()* function that allows the return object type to be pre-specified. Function *rapply()* is a recursive apply for nested lists, i.e. lists within lists. Function *tapply()* is a tagged apply where the tags identify the subsets. Function *apply()* is generic, applies a function to a matrix's rows or columns or, more generally, to dimensions of an array.

If we want to apply a function to the rows or columns of a matrix or array, we use the *apply()* function as below.

```
> M <- matrix(seq(1,16), 4, 4)
> M
     [,1] [,2] [,3] [,4]
[1,]    1    5    9   13
[2,]    2    6   10   14
[3,]    3    7   11   15
[4,]    4    8   12   16

> apply(M, 1, min)
[1] 1 2 3 4

> apply(M, 2, max)
[1] 4 8 12 16
> M <- array( seq(32), dim = c(4,4,2))
> M
, , 1
```

```
       [,1] [,2] [,3] [,4]
[1,]    1    5    9   13
[2,]    2    6   10   14
[3,]    3    7   11   15
[4,]    4    8   12   16
, , 2

       [,1] [,2] [,3] [,4]
[1,]   17   21   25   29
[2,]   18   22   26   30
[3,]   19   23   27   31
[4,]   20   24   28   32

> apply(M, 1, sum)
[1] 120 128 136 144

> apply(M, c(1,2), sum)
      [,1] [,2] [,3] [,4]
[1,]   18   26   34   42
[2,]   20   28   36   44
[3,]   22   30   38   46
[4,]   24   32   40   48
```

If we want to apply a function to each element of a list in turn and get a list back, we use the *lapply()* function as below.

```
> x <- list(a = 1, b = 1:3, c = 10:100)
> x
$a
[1] 1
$b
[1] 1 2 3
```

```
$c
 [1]  10  11  12  13  14  15  16  17  18  19  20  21  22  23  24  25  26
[18]  27  28  29  30  31  32  33  34  35  36  37  38  39  40  41  42  43
[35]  44  45  46  47  48  49  50  51  52  53  54  55  56  57  58  59  60
[52]  61  62  63  64  65  66  67  68  69  70  71  72  73  74  75  76  77
[69]  78  79  80  81  82  83  84  85  86  87  88  89  90  91  92  93  94
[86]  95  96  97  98  99 100

> lapply(x, FUN = length)
$a
[1] 1
$b
[1] 3
$c
[1] 91

> lapply(x, FUN = sum)
$a
[1] 1
$b
[1] 6
$c
[1] 5005
```

We use the function *sapply()*, if we want to apply a function to each element of a list in turn, and we want a vector back.

```
> x <- list(a = 1, b = 1:3, c = 10:100)
> x
$a
[1] 1
```

$b

[1] 1 2 3

$c

[1] 10 11 12 13 14 15 16 17 18 19 20 21 22 23 24 25 26

[18] 27 28 29 30 31 32 33 34 35 36 37 38 39 40 41 42 43

[35] 44 45 46 47 48 49 50 51 52 53 54 55 56 57 58 59 60

[52] 61 62 63 64 65 66 67 68 69 70 71 72 73 74 75 76 77

[69] 78 79 80 81 82 83 84 85 86 87 88 89 90 91 92 93 94

[86] 95 96 97 98 99 100

```
> sapply(x, FUN = length)
a      b      c
1      3      91
> sapply(x, FUN = sum)
a      b      c
1      6      5005
```

When we want to use the function *sapply()*, but need to squeeze some more speed out of the code, we use the function *vapply()* as below. For the function *vapply()*, we give R the information on what the function will return, which can save some time coercing returned values to fit in a single atomic vector. In the example below, we tell R that everything returned by *length()* should be an integer of length 1.

```
> x <- list(a = 1, b = 1:3, c = 10:100)
> vapply(x, FUN = length, FUN.VALUE = 0L)
a      b      c
1      3      91
```

For when we have several data structures (e.g. vectors, lists) and we want to apply a function to the 1st elements of each, and then the 2nd elements of each, etc., coercing the result to a vector/array we use the function *vapply()* as below.

```
> mapply(sum, 1:5, 1:5, 1:5)
[1] 3 6 9 12 15
> mapply(rep, 1:4, 4:1)
[[1]]
[1] 1 1 1 1
[[2]]
[1] 2 2 2
[[3]]
[1] 3 3
[[4]]
[1] 4
```

When we want to apply a function to each element of a nested list structure, recursively, we use the function *rapply()* as below. The function *rapply()* can be best illustrated with a user defined function to be applied.

```
> myFun <- function(x) {
+    if (is.character(x)) {
+       return(paste(x,"!",sep=""))
+    }
+    else {
+       return(x + 1)
+    }
+ }

> l <- list(a = list(a1 = "Boo", b1 = 2, c1 = "Eeek"),
+        b = 3, c = "Yikes",
+        d = list(a2 = 1, b2 = list(a3 = "Hey", b3 = 5)))

> rapply(l, myFun)
```

| a.a1 | a.b1 | a.c1 | b | c | d.a2 | d.b2.a3 | d.b2.b3 |
|------|------|------|---|---|------|---------|---------|
| "Boo!" | "3" | "Eeek!" | "4" | "Yikes!" | "2" | "Hey!" | "6" |

```
> rapply(l, myFun, how = "replace")
$a
$a$a1
[1] "Boo!"
$a$b1
[1] 3
$a$c1
[1] "Eeek!"
$b
[1] 4
$c
[1] "Yikes!"
$d
$d$a2
[1] 2
$d$b2
$d$b2$a3
[1] "Hey!"
$d$b2$b3
[1] 6
```

When we want to apply a function to subsets of a vector and the subsets are defined by some other vector, usually a factor, we use the function *tapply()* as below.

```
> x <- 1:20
> x
[1] 1 2 3 4 5 6 7 8 9 10 11 12 13 14 15 16 17 18 19 20
```

---

```
> y <- factor(rep(letters[1:5], each = 4))
> y
```

[1] a a a a b b b b c c c c d d d d e e e e

Levels: a b c d e

```
> tapply(x, y, sum)
```

| a | b | c | d | e |
|----|----|----|----|----|
| 10 | 26 | 42 | 58 | 74 |

The *by()* function, can be thought of, as a "wrapper" for the function *tapply()*. When we want to compute a task that *tapply()* can't handle, the *by()* function arises.

```
> cta <- tapply(iris$Sepal.Width , iris$Species , summary )
> cba <- by(iris$Sepal.Width , iris$Species , summary )
> cta
```

$setosa

| Min. | 1st Qu. | Median | Mean | 3rd Qu. | Max. |
|------|---------|--------|------|---------|------|
| 2.300 | 3.200 | 3.400 | 3.428 | 3.675 | 4.400 |

$versicolor

| Min. | 1st Qu. | Median | Mean | 3rd Qu. | Max. |
|------|---------|--------|------|---------|------|
| 2.000 | 2.525 | 2.800 | 2.770 | 3.000 | 3.400 |

$virginica

| Min. | 1st Qu. | Median | Mean | 3rd Qu. | Max. |
|------|---------|--------|------|---------|------|
| 2.200 | 2.800 | 3.000 | 2.974 | 3.175 | 3.800 |

```
> cba
```

iris$Species: setosa

| Min. | 1st Qu. | Median | Mean | 3rd Qu. | Max. |
|------|---------|--------|------|---------|------|
| 2.300 | 3.200 | 3.400 | 3.428 | 3.675 | 4.400 |

--------------------------------------------------------

*iris$Species: versicolor*

> Min. 1st Qu. Median   Mean 3rd Qu.   Max.
>
> 2.000  2.525  2.800  2.770  3.000  3.400

-----------------------------------------------------

*iris$Species: virginica*

> Min. 1st Qu. Median   Mean 3rd Qu.   Max.
>
> 2.200  2.800  3.000  2.974  3.175  3.800

If we print these two objects, *cta* and *cba*, we have the same results. The only differences are in how they are shown with the different class attributes. The power of the function *by()* arises when we can't use the function *tapply()* as in the following code.

> *tapply(iris, iris$Species, summary )*

*Error in tapply(iris, iris$Species, summary) :*

*arguments must have same length*

R says that arguments must have the same lengths, say "we want to calculate the summary of all variable in iris along the factor Species": but R just can't do that because it does not know how to handle. The *by()* function lets the *summary()* function work even if the length of the first argument are different.

> *bywork <- by(iris, iris$Species, summary )*

> *bywork*

*iris$Species: setosa*

> Sepal.Length  Sepal.Width   Petal.Length   Petal.Width
>
> Min.   :4.300  Min.   :2.300  Min.   :1.000  Min.   :0.100
>
> 1st Qu.:4.800  1st Qu.:3.200  1st Qu.:1.400  1st Qu.:0.200
>
> Median :5.000  Median :3.400  Median :1.500  Median :0.200
>
> Mean   :5.006  Mean   :3.428  Mean   :1.462  Mean   :0.246
>
> 3rd Qu.:5.200  3rd Qu.:3.675  3rd Qu.:1.575  3rd Qu.:0.300
>
> Max.   :5.800  Max.   :4.400  Max.   :1.900  Max.   :0.600

*Species*

*setosa   :50*

*versicolor: 0*

*virginica : 0*

-------------------------------------------------------

*iris$Species: versicolor*

*Sepal.Length   Sepal.Width   Petal.Length   Petal.Width*

*Min.   :4.900   Min.   :2.000   Min.   :3.00   Min.   :1.000*

*1st Qu.:5.600   1st Qu.:2.525   1st Qu.:4.00   1st Qu.:1.200*

*Median :5.900   Median :2.800   Median :4.35   Median :1.300*

*Mean   :5.936   Mean   :2.770   Mean   :4.26   Mean   :1.326*

*3rd Qu.:6.300   3rd Qu.:3.000   3rd Qu.:4.60   3rd Qu.:1.500*

*Max.   :7.000   Max.   :3.400   Max.   :5.10   Max.   :1.800*

*Species*

*setosa    : 0*

*versicolor:50*

*virginica : 0*

-------------------------------------------------------

*iris$Species: virginica*

*Sepal.Length   Sepal.Width   Petal.Length   Petal.Width*

*Min.   :4.900   Min.   :2.200   Min.   :4.500   Min.   :1.400*

*1st Qu.:6.225   1st Qu.:2.800   1st Qu.:5.100   1st Qu.:1.800*

*Median :6.500   Median :3.000   Median :5.550   Median :2.000*

*Mean   :6.588   Mean   :2.974   Mean   :5.552   Mean   :2.026*

*3rd Qu.:6.900   3rd Qu.:3.175   3rd Qu.:5.875   3rd Qu.:2.300*

*Max.   :7.900   Max.   :3.800   Max.   :6.900   Max.   :2.500*

*Species*

*setosa    : 0*

*versicolor: 0*

*virginica :50*

The arguments must have the same lengths. R can't do that because it does not know how to handle it. The *by()* function lets the *summary()* function work even if the length of the first argument is different. The result is an object of class by that along Species computes the summary of each variable.

The *aggregate()* function can be seen as another a different way of using *tapply()* function if we use it in such a way.

> *att <- tapply(iris$Sepal.Length , iris$Species , mean)*

> *agt <- aggregate(iris$Sepal.Length , list(iris$Species), mean)*

> *att*

   *setosa versicolor  virginica*

   *5.006    5.936    6.588*

> *agt*

|   | Group.1 | x |
|---|---------|-------|
| 1 | setosa | 5.006 |
| 2 | versicolor | 5.936 |
| 3 | virginica | 6.588 |

The two immediate differences are that the second argument of the *aggregate()* function must be a list while *tapply()* function can (not mandatory) be a list and that the output of the *aggregate()* function is a data frame while the one of *tapply()* function is an array. The power of the *aggregate()* function is that it can handle easily subsets of the data with *subset* argument and that it can handle formula as well. These elements make the *aggregate()* function easier to work with than *tapply()* function in some situations.

> *ag <- aggregate(len ~ ., data = ToothGrowth, mean)*

> *ag*

|   | supp | dose | len |
|---|------|------|-----|
| 1 | OJ | 0.5 | 13.23 |
| 2 | VC | 0.5 | 7.98 |
| 3 | OJ | 1.0 | 22.70 |
| 4 | VC | 1.0 | 16.77 |
| 5 | OJ | 2.0 | 26.06 |
| 6 | VC | 2.0 | 26.14 |

## ❖ HIGHLIGHTS

- One of the packages in R is *datasets* which is filled with example datasets.
- We can see all the datasets available in the loaded packages using the *data()* function.
- The *read.table()* function reads the CSV files and stores the result in a data frame.
- The *system.file()* function is used to locate files that are inside a package.
- Writing data from R into a file is done using the functions *write.table()* and *write.csv()*.
- If the file is unstructured, it is read using the function *readLines()*.
- The *writeLines()* function takes a text line and the file name as argument and writes the text to the file.
- The XML file can be imported using the function *xmlParse()* function.
- The function used to import the JSON file is *fromJSON()* and the function used to export the JSON file is *toJSON()*.
- Spreadsheets can be imported with the functions *read.xlsx()* and *read.xlsx2()*.
- To write to an excel file from R we use the function *write.xlsx2()*.
- The *read.ssd()* function is used to read SAS datasets.
- The *read.spss()* function is used to import the SPSS data files.
- The MATLAB binary data files can be read and written using the *readMat()* and *writeMat()* functions in the *R.matlab* package.

- R can connect to all DBMS like SQLite, MySQL, MariaDB, PostgreSQL and Oracle using the DBI package.
- The function *dbReadTable()* reads a table from the connected database.
- The functions *grep()* and *grepl()* are used to find a pattern in a given text.
- The functions *sub()* and *gsub()* are used to replace a pattern with another in a given text.
- The function *str_split()* is used to split a given text based on the pattern specified.
- The function *str_count()* can be used to count the number of occurrence of a given pattern in the given text.
- The function *str_replace()* can be used to replace the specified pattern with another pattern in the given text.
- The function *na.omit()* will remove the rows with missing values in a data frame.
- The function *sort()* sorts the given vector of numbers or strings.
- The function *order()* is the inverse of the *sort()* function.
- The *rank()* function lists the rank of the elements in a vector or a data frame.
- The package *sqldf* needs to be installed to manipulate the data frames or datasets using SQL.
- Changing the shape of a data frame is done using the functions *melt()* and *cast()*.
- R has many grouping functions such as *apply()*, *lapply()*, *sapply()*, *vapply()*, *mapply()*, *rapply()*, *tapply()*, *aggregate()* and *by()*.

# CHAPTER 4

# GRAPHICS USING R

❖ **OBJECTIVES**

On completion of this Chapter you will be able to:

- understand what is Exploratory Data Analysis
- introduce about the main graphical packages
- draw pie charts using R
- draw scatter plots using R
- draw line plots using R
- draw histograms using R
- draw box plots using R
- draw bar plots using R
- know about other existing graphical packages

## 4.1. Exploratory Data Analysis

Exploratory Data Analysis (EDA) is a visual based method used to analyse data sets and to summarize their main characteristics. Exploratory Data Analysis (EDA) shows how to use visualisation and transformation to explore data in a systematic way. EDA is an iterative cycle of the below steps:

1) Generate questions about data.
2) Search for answers by visualising, transforming, and modelling data.
3) Use what is learnt to refine questions and/or generate new questions.

Exploratory Data Analysis (EDA) is an approach for data analysis that employs a variety of techniques (mostly graphical) to:

1)   Maximize insight into a data set
2)   Uncover underlying structure
3)   Extract important variables
4)   Detect outliers and anomalies
5)   Test underlying assumptions
6)   Develop parsimonious models
7)   Determine optimal factor settings.

## 4.2. Main Graphical Packages

The basic graphs in R can be drawn using the *base* graphics system. These have some limitations and they are overcome in the next level of graphics called the *grid* graphics system. This system allows to plot the points or lines in the place where desired. But, this does not allow us to draw a scatter plot. Hence, we go for the next level of plotting which the *lattice* graphics system is. In this system, the results of a plot can be saved. Also these scatter plots can contain multiple panels in which we can draw multiple graphs and compare them to each other. The next levels of graphs are the *ggplot2* graphics system. In this the *"gg"* stands for *"grammar of graphics"*. This breaks down the graphs into many parts or chunks.

## 4.3. Pie Charts

In R the pie chart is created using the *pie()* function which takes positive numbers as vector input. The additional parameters are used to control labels, colour, title etc. The basic syntax for creating a pie-chart is as given below and the explanation of the parameters are also listed.

*pie(x, labels, radius, main, col, clockwise)*

*x – numeric vector*

*labels – description of the slices*

*radius – values between [-1 to +1]*

*main – title of the chart*

*col – colour palette*

*clockwise – logical value – TRUE (Clockwise), FALSE (Anti Clockwise)*

> x <- c(25, 35, 10, 5, 15)

> labels <- c("Rose", "Lotus", "Lilly", "Sunflower", "Jasmine")

> pie(x, labels = percent, main = "Flowers", col = rainbow(length(x)))

> legend("topright", c("Rose", "Lotus", "Lilly", "Sunflower", "Jasmine"),

cex = 0.8, fill = rainbow(length(x)))

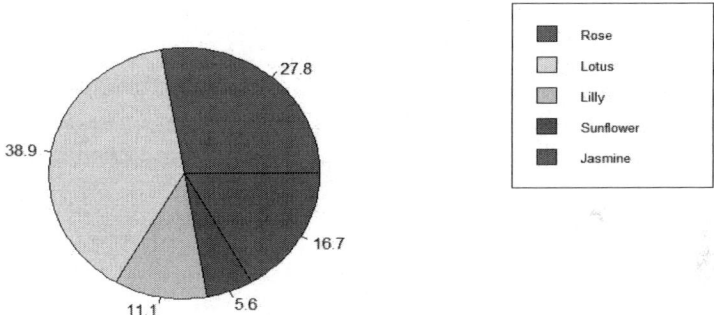

**Figure 4.1**   Pie Chart of Flowers

A 3D Pie Chart can be drawn using the package *plotrix* which uses the function *pie3D()*.

> install.packages("plotrix")

> library(plotrix)

> pie3D(x, labels = labels, explode = 0.1, main = "Flowers")

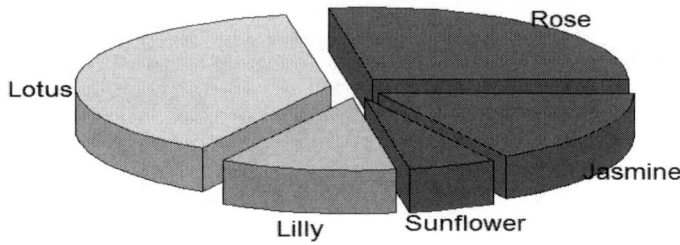

**Figure 4.2**  3-D Pie Chart of Flowers

## 4.4. Scatter Plots

Scatter plots are used for exploring the relationship between the two continuous variables. Let us consider the dataset *"cars"* that lists the "Speed and Stopping Distances of Cars". The basic scatter plot in the *base* graphics system can be obtained by using the *plot()* function as in *Fig. 4.3*. The below example compares if the speed of a car has effect on its stopping distance using the plot.

> *colnames(cars)*

*[1] "speed" "dist"*

> *plot(cars$speed, cars$dist)*

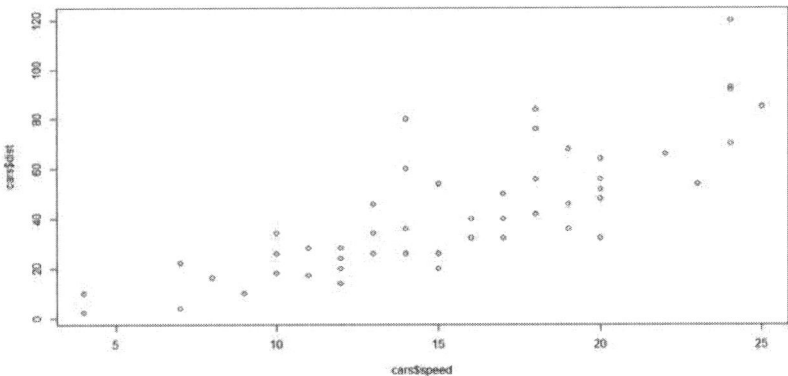

**Figure 4.3**   Basic Scatter Plot of Car Speed Vs Distance

This plot can be made more appealing and readable by adding colour and changing the plotting character. For this we use the arguments *col* and *pch* (can take the values between 1 and 25) in the *plot()* function as below. Thus the plot in *Fig. 4.4* shows that there is a strong positive correlation between the speed of a car and its stopping distance.

> *plot(cars$speed, cars$dist, col = "red", pch = 15)*

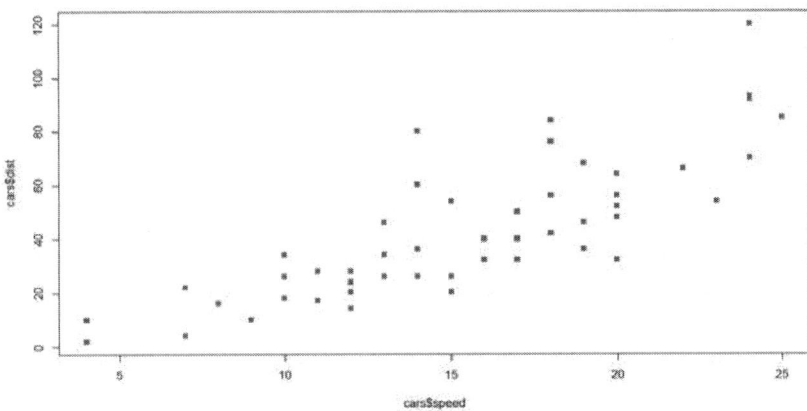

**Figure 4.4**   Coloured Scatter Plot of Car Speed Vs Distance

The *layout()* function is used to control the layout of multiple plots in the matrix. Thus in the example below multiple related plots are placed in a single figure as in *Fig. 4.5*.

> *data(mtcars)*

> *layout(matrix(c(1,2,3,4), 2, 2, byrow = TRUE))*

> *plot(mtcars$wt, mtcars$mpg, col = "blue", pch = 17)*

> *plot(mtcars$wt, mtcars$disp, col = "red", pch = 15)*

> *plot(mtcars$mpg, mtcars$disp, col = "dark green", pch = 10)*

> *plot(mtcars$mpg, mtcars$hp, col = "violet", pch = 7)*

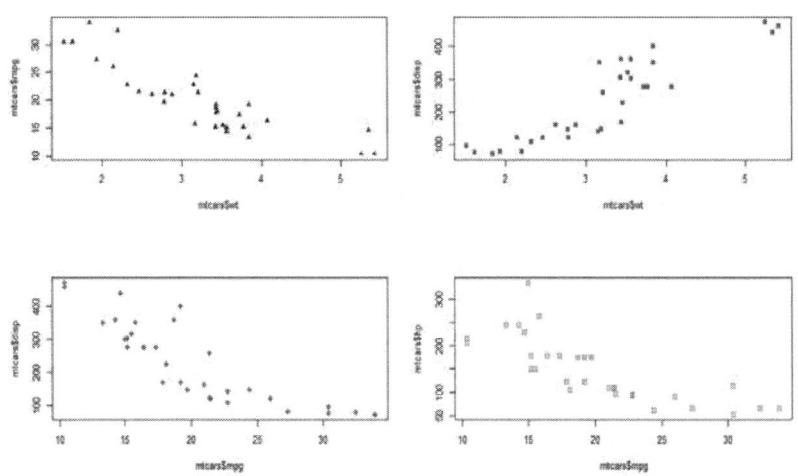

**Figure 4.5**   Layout of Multiple Scatter Plots

When we have more than two variables and we want to find the correlation between one variable versus the remaining ones we use scatter plot matrix. We use *pairs()* function to create matrices of scatter plots as in *Fig. 4.6*. The basic syntax for creating scatter plot matrices in R is as below.

*pairs(formula, data)*

> *pairs(~wt+mpg+disp+cyl,data = mtcars, main = "Scatterplot Matrix")*

**Scatterplot Matrix**

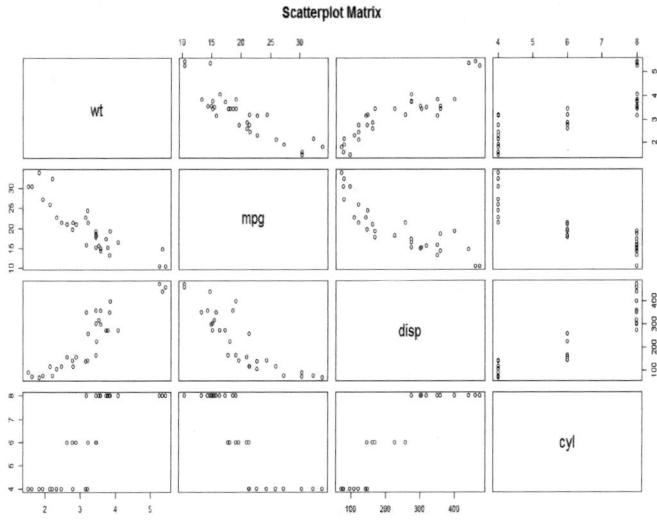

**Figure 4.6**   Scatter Plot Matrix Using pairs()

The *lattice* graphics system has equivalent of *plot()* function and it is *xyplot()*. This function uses a formula to specify the x and y variables (*yvar ~ xvar*) and a data frame argument. To use this function, it is required to include the *lattice* package.

> *library(lattice)*

> *xyplot(mtcars$mpg ~ mtcars$disp, mtcars, col = "purple", pch = 7)*

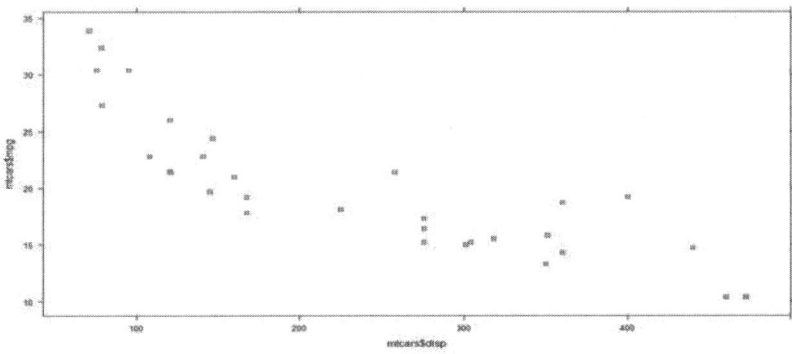

**Figure 4.7**   Scatter Plot Matrix Using xyplot()

Axis scales can be specified in the *xyplot()* using the *scales* argument and this argument must be a list. This list consists of the *name* = *value* pairs. If we mention *log* = *TRUE*, the log scales for the x and y axis are set as in *Fig. 4.8*. The scales list can take other arguments also like the *x* and *y* that sets the x and y axes respectively.

> *xyplot(mtcars$mpg ~ mtcars$disp, mtcars, scales = list(log = TRUE),*

> *col = "red", pch = 11)*

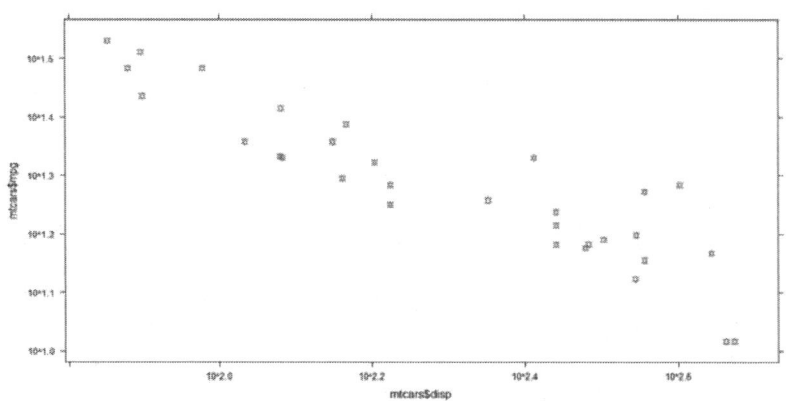

**Figure 4.8**   Scatter Plot Matrix with Axis Scales Using xyplot()

The data in the graph can be split based on one of the columns in the dataset namely *mtcars$carb*. This can be done by appending the pipe symbol ( | ) along with the column name used for splitting. The argument *relation* = *"same"* means that each panel shares the same axes. If the argument *alternating* = *TRUE*, axis ticks for each panel is drawn on alternating sides of the plot as in *Fig. 4.9*.

> *xyplot(mtcars$mpg ~ mtcars$disp | mtcars$carb, mtcars,*

> *scales = list(log = TRUE, relation = "same", alternating = FALSE),*

> *layout = c(3, 2), col = "blue", pch = 14)*

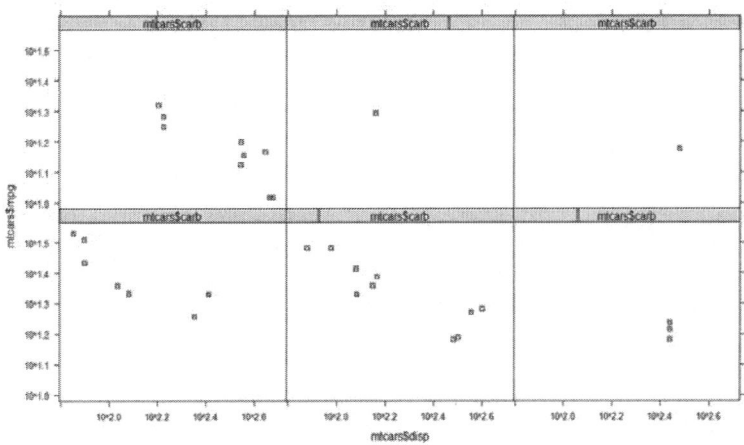

**Figure 4.9**   Scatter Plot Split on a Column

The lattice plots can be stored in variables and hence they can be further updated using the function update as below.

> *graph1 <- xyplot(mtcars$mpg ~ mtcars$disp | mtcars$carb, mtcars,*

> *scales = list(log = TRUE, relation = "same", alternating = FALSE),*

> *layout = c(3, 2), col = "blue", pch = 14)*

> *graph2 <- update(graph1, col = "yellow", pch = 6)*

In the *ggplot2* graphics, each plot is drawn with a call to the *ggplot()* function as in *Fig. 4.10*. This function takes a data frame as its first argument. The passing of data frame columns to the x and y axis is done using the *aes()* function which is used within the *ggplot()* function. The other aesthetics to the graph are then added using the *geom()* function appended with a "+" symbol to the *ggplot()* function.

> *library(ggplot2)*

> *ggplot(mtcars, aes(mpg, disp)) +*

> *geom_point(color = "purple", shape = 16, cex = 2.5)*

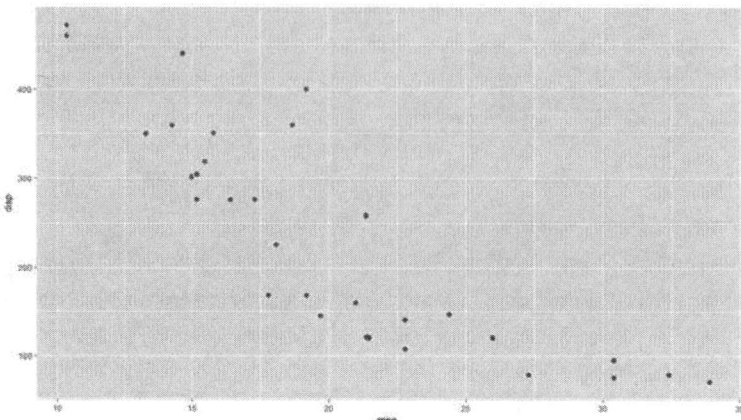

**Figure 4.10**   Scatter Plot Using ggplot()

The *ggplots* can also be split into several panels like the lattice plots as in *Fig. 4.11*. This is done using the function *facet_wrap()* which takes a formula of the column used for splitting. The function *theme()* is used to specify the orientation of the axis readings. The functions *facet_wrap()* and *theme()* are appended to the *ggplot()* function using the "+" symbol. The *ggplots* can be stored in a variable like the lattice plots and as usual wrapping the expression in parentheses makes it to auto print.

> *(graph1 <- ggplot(mtcars, aes(mpg, disp)) +*

   *geom_point(color = "dark green", shape = 15, cex = 3))*

> *(graph2 <- graph1 + facet_wrap(~mtcars$cyl, ncol = 3) +*

   *theme(axis.text.x = element_text(angle = 90, hjust = 1)*

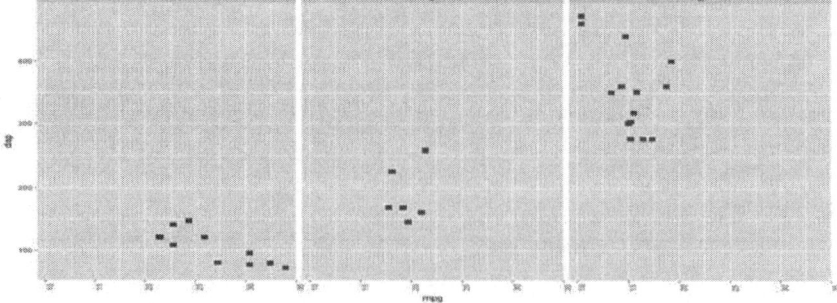

**Figure 4.11 – Scatter Plot Split into Panels Using ggplot()**

## 4.5. Line Plots

A line chart / line plot is a graph that connects a series of points by drawing line segments between them. Line charts are usually used in identifying the trends in data. The *plot()* function in R is used to create the line graph in *base* graphics as in *Fig. 4.12*. This function takes a vector of numbers as input together with few more parameters listed below.

*plot(v, type, col, xlab, ylab)*

*v – numeric vector*

*type - takes value "p" (only points), or "l" (only lines) or "o" (both points and lines)*

*xlab – label of x-axis*

*ylab – label of y-axis*

*main - title of the chart*

*col - colour palette*

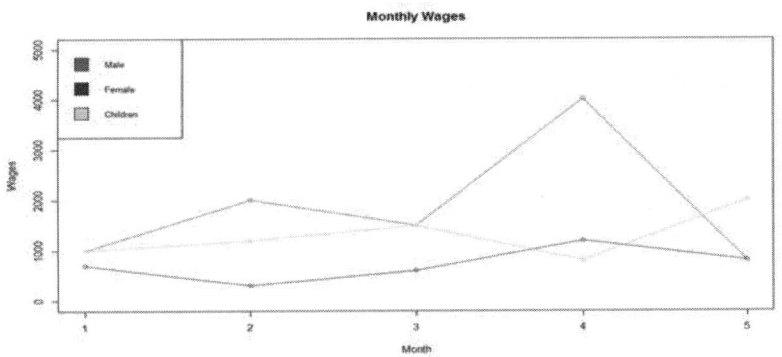

**Figure 4.12** Line Plot Using Basic Graphics

> *male <- c(1000, 2000, 1500, 4000, 800)*

> *female <- c(700, 300, 600, 1200, 800)*

> *child <- c(1000, 1200, 1500, 800, 2000)*

> *wages <- c("Male", "Female", "Children")*

> *color = c("red", "blue", "green")*

> plot(male, type = "o", col = "red", xlab = "Month", ylab = "Wages",

> main = "Monthly Wages", ylim = c(0, 5000))

> lines(female, type = "o", col = "blue")

> lines(child, type = "o", col = "green")

> legend("topleft", wages, cex = 0.8, fill = color)

Line plots in the *lattice* graphics uses the *xyplot()* function as in *Fig. 4.13*. In this multiple lines can be creating using the "+" symbol in the formula where the x and the y axes are mentioned. The argument *type* = "*l*" is used to mention that it is a continuous line.

> xyplot(economics$pop + economics$unemploy ~ economics$date, economics, type = "l")

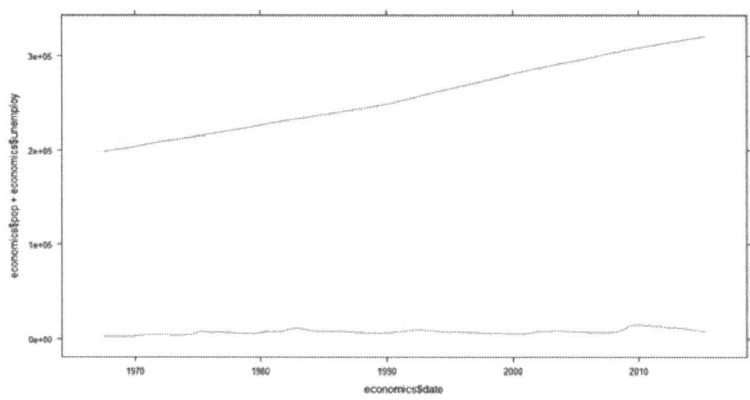

**Figure 4.13**　Line Plot Using Lattice Graphics

In the *ggplot2* graphics, the same syntax for scatter plots are used, except for the change of *geom_plot()* function with the *geom_line()* function as in *Fig. 4.14*. But, there need to be multiple *geom_line()* functions for multiple lines to be drawn in the graph.

> ggplot(economics, aes(economics$date)) + geom_line(aes(y = economics$pop)) +

> geom_line(aes(y = economics$unemploy))

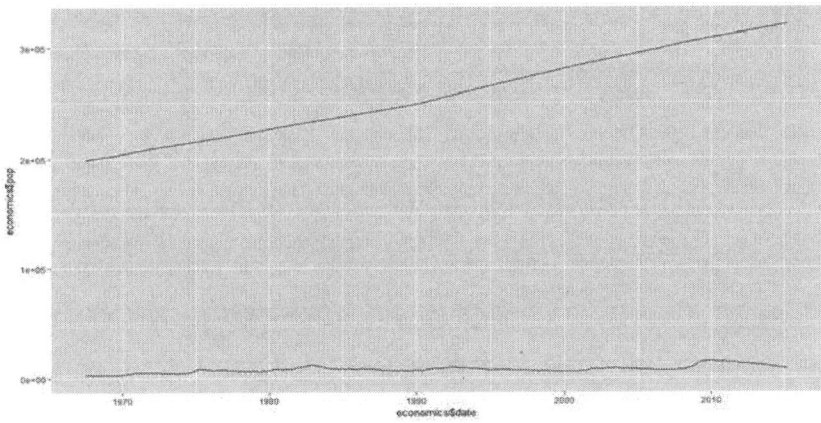

**Figure 4.14**   Line Plot Using ggplot2 Graphics

The plot in the Fig. 4.15 can be drawn without using multiple *geom_line()* functions also. This is possible using the function *geom_ribbon()* as mentioned below. This function plots not only the two lines, but also the contents in between the two lines.

> *ggplot(economics, aes(economics$date, ymin = economics$unemploy,*

   *ymax = economics$pop)) + geom_ribbon(color = "blue", fill = "white")*

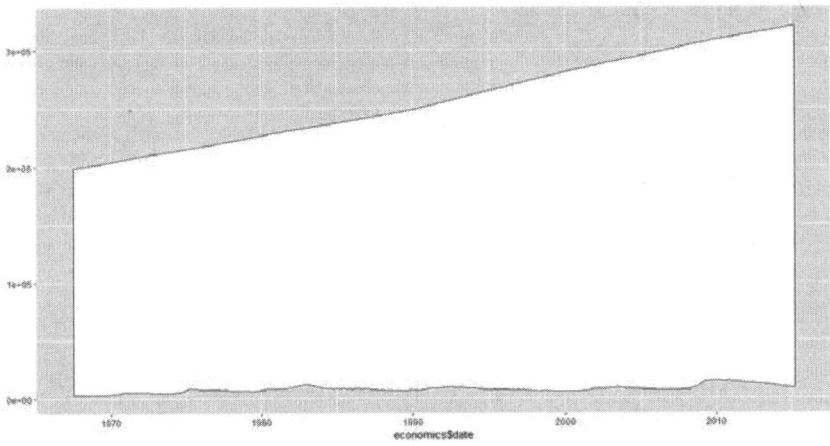

**Figure 4.15**   Line Plot Using geom_ribbon()

## 4.6. Histograms

Histograms represents the variable values frequencies, that are split into ranges. This is similar to bar charts, but histograms group values into continuous ranges. In R histograms in the *base* graphics are drawn using the function *hist()* as in the *Fig. 4.16*, that takes a vector of numbers as input together with few more parameters listed below.

*hist(v, main, xlab, xlim, ylim, breaks, col, border)*

| | |
|---|---|
| *v – numeric vector* | *main - title of the chart* |
| *col - colour palette* | *border – border colour* |
| *xlab – label of x-axis* | *xlim – range of x-axis* |
| *ylim – range of y-axis* | *breaks – width of each bar* |

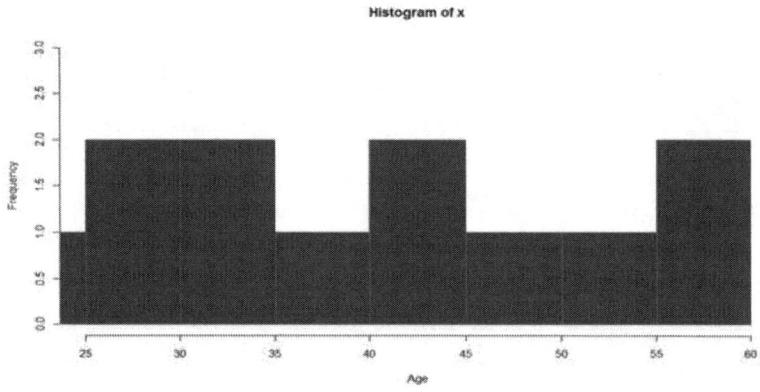

**Figure 4.16**   Histogram Using Base Graphics

> x <- c(45, 33, 31, 23, 58, 47, 39, 58, 28, 55, 42, 27)

> hist(x, xlab = "Age", col = "blue", border = "red", xlim = c(25, 60),

ylim = c(0, 3), breaks = 5)

The *lattice* histogram is drawn using the function *histogram()* as in *Fig. 4.17* and it behaves in the same way as the base ones. But it allows easy splitting of data into panels and saving plots as variables. The *breaks* argument behaves the same way as with *hist()*. The *lattice* histograms support counts, probability densities, and percentage y-axes via the *type* argument, which takes the string *"count"*, *"density"*, or *"percent"*.

> *histogram(~ mtcars$mpg, mtcars, breaks = 10)*

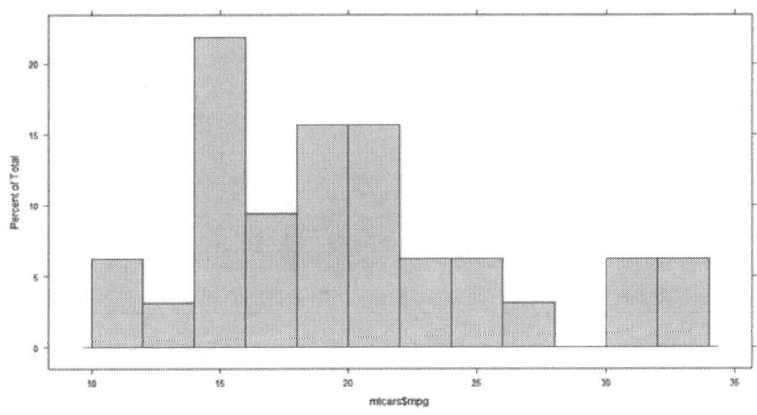

**Figure 4.17**   Histogram Using Lattice Graphics

The *ggplot* histograms are created by adding the function *geom_histogram()* to the *ggplot()* function as in *Fig. 4.18*. Bin specification is simple here, we just need to pass a numeric bin width to *geom_histogram()* function. It is possible to choose between counts and densities by passing the special names *..count..* or *..density..* to the y-aesthetic.

> *ggplot(mtcars, aes(mtcars$mpg, ..density..)) + geom_histogram(binwidth = 5)*

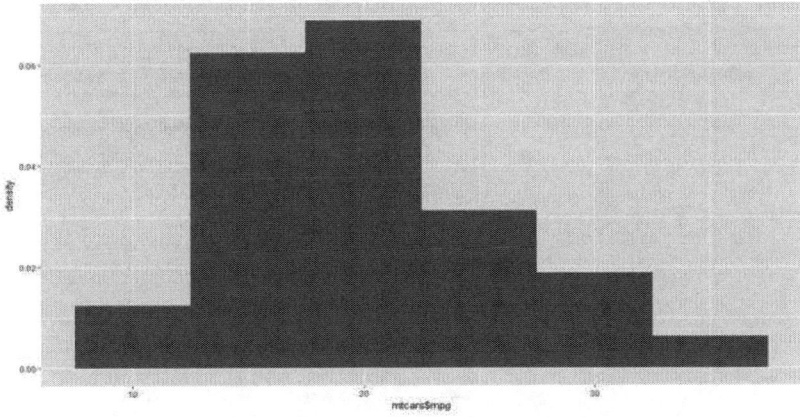

**Figure 4.18**   Histogram Using ggplot2 Graphics

## 4.7. Box Plots

The box plot divides the data into three quartiles. This graph represents the minimum, maximum, median, first quartile and third quartile in the data. This shows the data distribution by drawing the box plots. In R *base* graphics the box plot is created using the *boxplot()* function as in *Fig. 4.19*, which takes the following parameters. The parameters are used to give the data as a data frame, a vector or a formula, a logical value to draw a notch, a logical value to draw a box as per the width of the sample, give title of the chart, labels for the boxes. The basic syntax for creating a box-plot is as given below and the explanation of the parameters are also listed.

*boxplot(x, data, notch, varwidth, names, main)*

*x – vector or a formula*

*data – data frame*

*notch – logical value (TRUE – draw a notch)*

*varwidth – logical value (TRUE – box width proportionate to sample size*

*names – labels printed under the boxes*

*main – title of the chart*

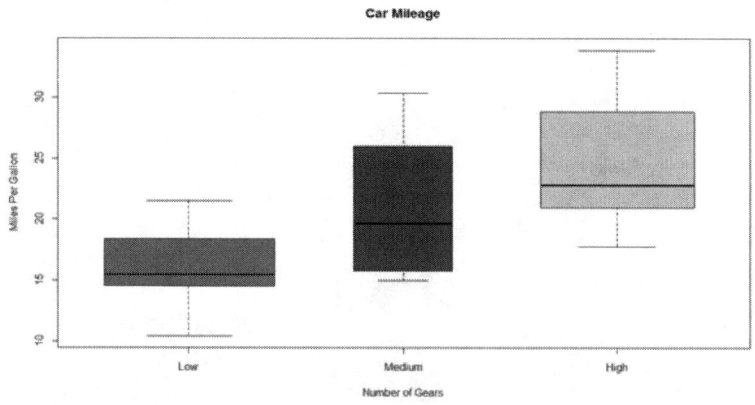

**Figure 4.19**  Box Plots Using Base Graphics

This type of plot is often clearer if we reorder the box plots from smallest to largest, in some sense. The *reorder()* function changes the order of a factor's levels, based upon some numeric score.

> *boxplot(mpg ~ reorder(gear, mpg, median), data = mtcars,*

  *xlab = "Number of Gears", ylab = "Miles Per Gallon",*

  *main = "Car Mileage", varwidth = TRUE,*

  *col = c("red","blue", "green"), names = c("Low", "Medium", "High"))*

In the *lattice* graphics the box plot is drawn using the function *bwplot()* as in Fig. 4.20.

> *bwplot(mpg ~ reorder(gear, mpg, median), data = mtcars,*

  *xlab = "Number of Gears", ylab = "Miles Per Gallon",*

  *main = "Car Mileage", varwidth = TRUE,*

  *col = c("red","blue", "green"), names = c("Low", "Medium", "High"))*

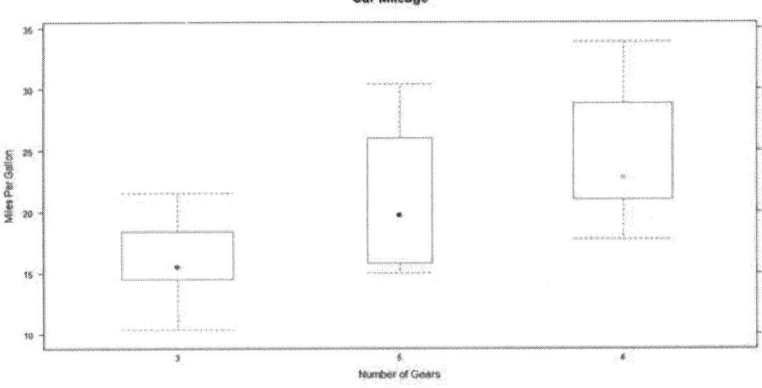

**Figure 4.20**   Box Plots Using Lattice Graphics

In the *ggplot2* graphics the box plot is drawn by adding the function *geom_boxplot()* to the function *ggplot()* as in Fig. 4.21.

> *ggplot(mtcars, aes(reorder(gear, mpg, median), mpg)) + geom_boxplot()*

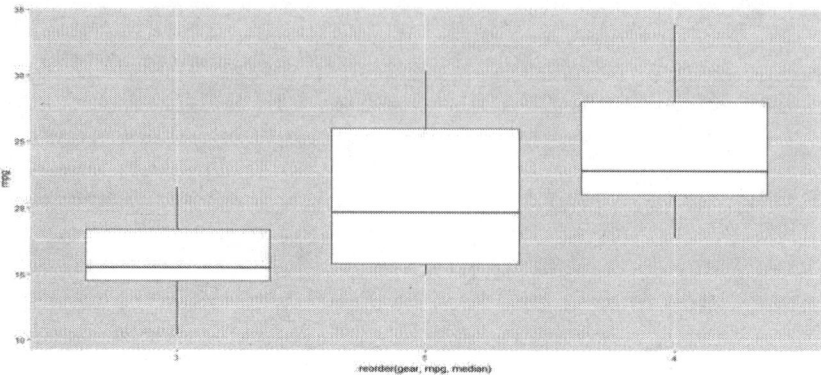

**Figure 4.21**    Box Plots Using ggplot2 Graphics

## 4.8. Bar Plots

Bar charts are the natural way of displaying numeric variables split by a categorical variable. In R *base* graphics the bar chart is created using the *barplot()* function as in *Fig. 4.22*, which takes a matrix or a vector of numeric values. The additional parameters are used to give labels to the X-axis, Y-axis, give title of the chart, labels for the bars and colours. The basic syntax for creating a bar-chart is as given below and the explanation of the parameters are also listed.

*barplot(H, xlab, ylab, main, names.arg, col)*

*H – numeric vector or matrix*

*x-lab – label of x-axis*

*y-lab – label of y-axis*

*main - title of the chart*

*names.arg – vector of labels under each bar*

*col – colour palette*

*> x <- matrix(c(1000, 900, 1500, 4400, 800, 2100, 1700, 2900, 3800),*

*nrow = 3, ncol = 3)*

*> years <- c("2011", "2012", "2013")*

*> city <- c("Chennai", "Mumbai", "Kolkata")*

> color <- c("red", "blue", "green")

> barplot(x, main = "Yearly Sales", names.arg = years, xlab = "Year",

ylab = "Sales", col = color)

> legend("topleft", city, cex = 0.8, fill = color)

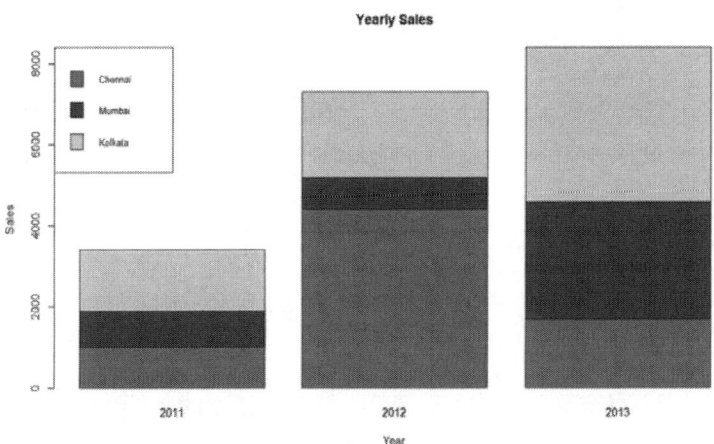

**Figure 4.22**   Vertical Bar Plot Using Base Graphics

By default the bars are vertical, but if we want horizontal bars, they can be generated with *horiz = TRUE* parameter as in *Fig. 4.23*. We can also do some fiddling with the plot parameters, via the *par()* function. The *las* parameter controls whether labels are horizontal, vertical, parallel, or perpendicular to the axes. Plots are usually more readable if you set *las = 1*, for horizontal. The *mar* parameter is a numeric vector of length 4, giving the width of the plot margins at the bottom/left/top/right of the plot.

> x <- matrix(c(1000, 900, 1500, 4400, 800, 2100, 1700, 2900, 3800), nrow = 3, ncol = 3)

> years <- c("2011", "2012", "2013")

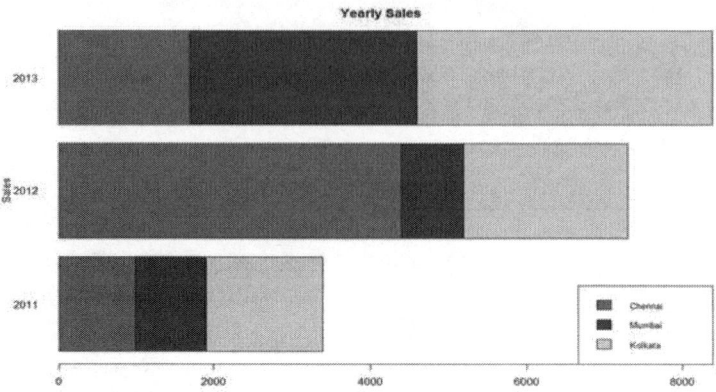

**Figure 4.23**  Horizontal Bar Plot Using Base Graphics

> *city <- c("Chennai", "Mumbai", "Kolkata")*

> *color <- c("red", "blue", "green")*

> *par(las = 1, mar = c(3, 9, 1, 1))*

> *barplot(x, main = "Yearly Sales", names.arg = years,*

>         *xlab = "Year", ylab = "Sales", col = color, horiz = TRUE)*

> *legend("bottomright", city, cex = 0.8, fill = color)*

The *lattice* equivalent of the function *barplot()*, is the function *barchart()* as shown in *Fig. 4.24*. The formula interface is the same as those we saw with scatter plots, *yvar ~ xvar*.

> *barchart(mtcars$mpg ~ mtcars$disp, mtcars)*

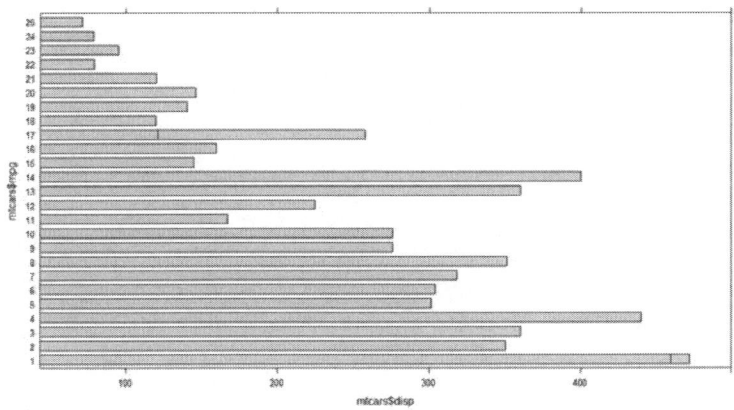

**Figure 4.24**   Horizontal Bar Plot Using Lattice Graphics

Extending this to multiple variables just requires a tweak to the formula, and passing *stack = TRUE* to make a stacked plot as in *Fig. 4.25*.

> *barchart(mtcars$mpg ~ mtcars$disp + mtcars$qsec + mtcars$hp, mtcars,*

*stack = TRUE*

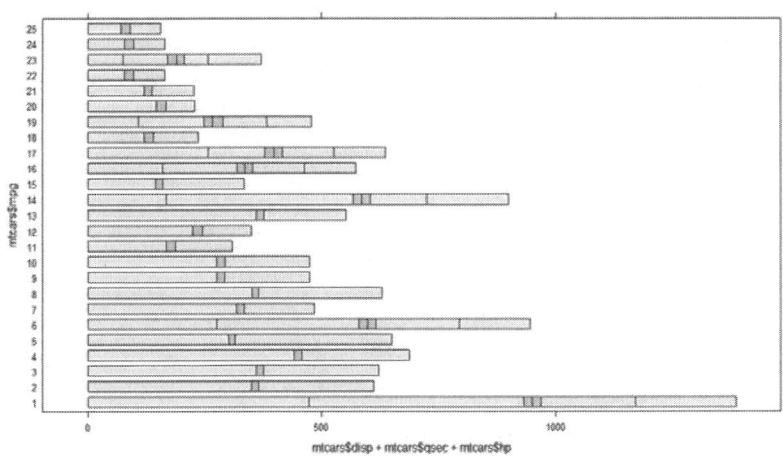

**Figure 4.25**   Horizontal Stacked Bar Plot Using Lattice Graphics

In the *ggplot2* graphics the bar chart is drawn by adding the function *geom_bar()* to the function *ggplot()* as in *Fig. 4.26*. Like *base*, *ggplot2* defaults to vertical bars; adding the function *coord_flip()* swaps this. We must pass the argument *stat = "identity"* to the function *geom_bar()*.

> *ggplot(mtcars, aes(mtcars$mpg, mtcars$disp)) + geom_bar(stat = "identity") +*

*coord_flip()*

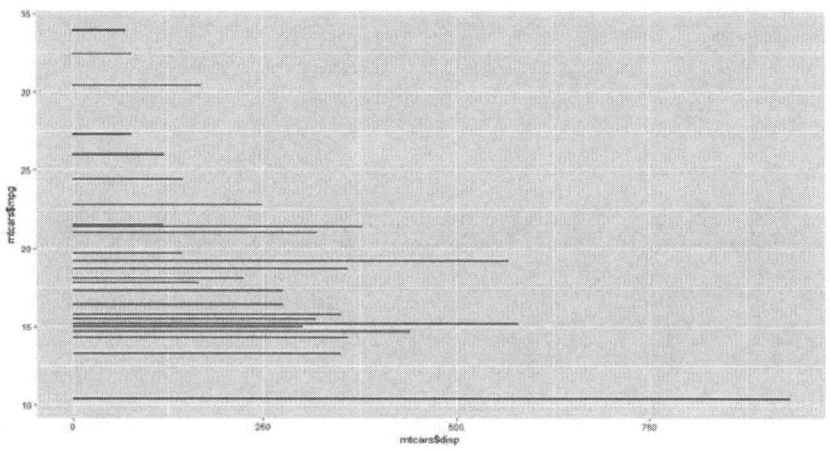

**Figure 4.26**   Horizontal Bar Plot Using ggplot2 Graphics

## 4.9. Other Graphical Packages

| Package Name | Description |
|---|---|
| Vcd | Plots for visualizing categorical data, such as mosaic plots and association plots |
| Plotrix | Loads of extra plot types |
| latticeExtra | Extends the *lattice* package |
| GGally | Extend the *ggplot2* package |
| Grid | Provide access to the underlying framework of lattice and ggplot2 packages |

| Package Name | Description |
| --- | --- |
| gridSVG | Write grid-based plots to SVG files |
| Playwith | Allows pointing and clicking to interact with base or lattice plots |
| Iplots | Provides a whole extra system of plots with more interactivity |
| googleVis | Provides an R wrapper around Google Chart Tools, creating plots that can be displayed in a browser. |
| Rggobi | Provides an interface to GGobi package (for visualizing high-dimensional data) |
| GGobi | Open source visualization program for exploring high-dimensional data. |
| Rgl | Provides an interface to OpenGL for interactive 3D plots |
| Animation | Lets to make animated GIFs or SWF animations |
| rCharts | Provides wrappers to half a dozen JavaScript plotting libraries using lattice syntax. |

## ❖ HIGHLIGHTS

- Exploratory Data Analysis (*EDA*) shows how to use visualisation and transformation to explore data in a systematic way.
- The main graphical packages are *base, lattice* and *ggplot2.*
- In R the pie chart is created using the *pie()* function.
- A 3D Pie Chart can be drawn using the package plotrix which uses the function *pie3D().*
- The basic scatter plot in the base graphics system can be obtained by using the *plot()* function.
- We use the arguments *col* and *pch* (values between 1 and 25) in the *plot()* function to specify colour and plot pattern.
- The *layout()* function is used to control the layout of multiple plots in the matrix.

- We use *pairs()* function to create matrices of scatter plots.
- The *lattice* graphics system has equivalent of *plot()* function and it is *xyplot()*.
- In the *ggplot2* graphics, each plot is drawn with a call to the *ggplot()* function.
- The *ggplots* can also be split into several panels like the *lattice* plots and this is done using the function *facet_wrap()*.
- The *plot()* function in R is used to create the line graph in base graphics and the argument *type* = "*l*" is used to mention that it is a line.
- Line plots in the *lattice* graphics uses the *xyplot()* function and the argument *type* = "*l*" is used to mention that it is a line.
- The *ggplot2* scatter plot are created by adding the function *geom_line()* to the *ggplot()* function.
- In R histograms in the *base* graphics are drawn using the function *hist()*.
- The *lattice* histogram is drawn using the function *histogram()*.
- The *ggplot2* histograms are created by adding the function *geom_histogram()* to the *ggplot()* function.
- In R *base* graphics the box plot is created using the *boxplot()* function.
- In the *lattice* graphics the box plot is drawn using the function *bwplot()*.
- In the *ggplot2* graphics the box plot is drawn by adding the function *geom_boxplot()* to the function *ggplot()*.
- In R base graphics the bar chart is created using the *barplot()* function.
- The lattice equivalent of the function *barplot()*, is the function *barchart()*.
- In the *ggplot2* graphics the bar chart is drawn by adding the function *geom_bar()* to the function *ggplot()*.

# CHAPTER 5

# STATISTICAL ANALYSIS USING R

## ❖ OBJECTIVES

On completion of this Chapter you will be able to:

- obtain the basic statistical measures like mean, median, mode, standard deviation, variation etc., using R
- obtain the summary statistics of a given data
- understand and plot the normal distribution of data using R functions
- understand and plot the binomial distribution of data using R functions
- perform correlation analysis on the given data using R
- perform regression analysis on the given data using R
- perform ANOVA, ANCOVA on the given data using R
- perform chi-square and hypothesis testing on the given data using R

Statistical analysis in R is performed by using many in-built functions. Most of these functions are part of the R *base* package. These functions take R vector as an input along with the arguments and give the result. The other important R package for statistical analysis is the *stats* package.

## 5.1. Basic Statistical Measures

Any dataset available in R or that is been imported into R for further analysis will have both categorical data as well as numeric data. So, we can apply the statistical functions available in R on the numeric data and understand the statistical measures of the fields. The basic statistical measures are the minimum, maximum, mean and median represented by the functions *min()*, *max()*, *mean()* and *median()*

respectively. Let us use the dataset named *mtcars* that is available in R by default to understand these statistical measures.

> *data(mtcars)*

> *colnames(mtcars)*

*[1] "mpg" "cyl" "disp" "hp"  "drat" "wt"  "qsec" "vs"  "am"  "gear" "carb"*

> *min(mtcars$cyl)*

*[1] 4*

> *max(mtcars$cyl)*

*[1] 8*

> *mean(mtcars$cyl)*

*[1] 6.1875*

> *median(mtcars$cyl)*

*[1] 6*

All the above results can also be obtained by one function *summary()* and this can also be applied on all the fields of the dataset at one shot. The *range()* function gives the minimum and maximum values of a numeric field at one go.

> *summary(mtcars$cyl)*

   *Min. 1st Qu.  Median   Mean 3rd Qu.   Max.*

   *4.000  4.000  6.000  6.188  8.000  8.000*

> *range(mtcars$cyl)*

*[1] 4 8*

## 5.1.1. Mean

Mean is calculated by taking the sum of the values and dividing with the number of values in a data series. The function *mean()* is used to calculate this in R. The basic syntax for calculating mean in R is given below along with its parameters.

*mean(x, trim = 0, na.rm = FALSE, ...)*

*x - numeric vector*

*trim - to drop some observations from both end of the sorted vector*

*na.rm - to remove the missing values from the input vector*

> *x <- c(45, 56, 78, 12, 3, -91, -45, 15, 1, 24)*

> *mean(x)*

*[1] 9.8*

When *trim* parameter is supplied, the values in the vector get sorted and then the required numbers of observations are dropped from calculating the mean. When trim = 0.2, 2 values from each end will be dropped from the calculations to find mean. In this case the sorted vector is (-91, -45, 1, 3, 12, 15, 24, 45, 56, 78) and the values removed from the vector for calculating mean are $(-91, -45)$ from left and (56, 78) from right.

> *mean(x, trim = 0.2)*

*[1] 16.66667*

If there are missing values, then the *mean()* function returns NA. To drop the missing values from the calculation use *na.rm = TRUE*, which means remove the NA values.

> *x <- c(45, 56, 78, 12, 3, -91, NA, -45, 15, 1, 24, NA)*

> *mean(x)*

*[1] NA*

> *mean(x, na.rm = TRUE)*

*[1] 9.8*

## 5.1.2. Median

The middle most value in a data series is called the median. The *median()* function is used in R to calculate this value. The basic syntax for calculating median in R is given below along with its parameters.

*median(x, na.rm = FALSE)*

*x - numeric vector*

*na.rm - to remove the missing values from the input vector*

```
> x <- c(45, 56, 78, 12, 3, -91, -45, 15, 1, 24)
> median(x)
[1] 13.5
```

## 5.1.3. Mode

The mode is the value that has highest number of occurrences in a set of data. Unlike mean and median, mode can have both numeric and character data. R does not have a standard in-built function to calculate mode. So we create a user function to calculate mode of a data set in R. This function takes the vector as input and gives the mode value as output.

```
Mode <- function(x)
{
 y <- unique(x)
 y[which.max(tabulate(match(x, y)))]
}

> x <- c(1,2,3,4,5,5,5)
> Mode(x)
[1] 5
> ch <- c("a", "e", "i", "o", "u", "u", "a", "a")
> Mode(ch)
[1] "a"
```

The function *unique()* returns a vector, data frame or array like *x* but with duplicate elements/rows removed. The function *match()* returns a vector of the positions of (first) matches of its first argument in its second. The function *tabulate()* takes the integer-valued vector bin and counts the number of times each integer occurs in it. The function *which.max()* determines the location, i.e., index of the (first) maximum of a numeric (or logical) vector.

## 5.1.4. Standard Deviation and Variance

The functions to calculate the standard deviation, variance and the mean absolute deviation are *sd()*, *var()* and *mad()* respectively.

> sd(mtcars$cyl)

[1] 1.785922

> var(mtcars$cyl)

[1] 3.189516

> mad(mtcars$cyl)

[1] 2.9652

## 5.1.5. Quartile Ranges

The *quantile()* function provides the quartiles of the numeric values. An alternative function for quartiles is *fivenum()*. The *IQR()* function provides the inter quartile range of the numeric fields.

> quantile(mtcars$cyl)

 0%  25%  50%  75% 100%

  4   4    6    8    8

> fivenum(mtcars$cyl)

[1] 4    4    6    8    8

> IQR(mtcars$cyl)

[1] 4

## 5.1.6. Other Statistical Functions

The function *cor()* and *cov()* are used to find the correlation and covariance between two numeric fields respectively. In the below example the value shows that there is negative correlation between the two numeric fields.

> cor(mtcars$mpg, mtcars$cyl)

[1] -0.852162

> *cov(mtcars$mpg, mtcars$cyl)*

*[1] -9.172379*

There are other statistics functions such as *pmin()*, *pmax()* [parallel equivalents of *min()* and *max()* respectively], *cummin()* [cumulative minimum value], *cummax()* [cumulative maximum value], *cumsum()* [cumulative sum] and *cumprod()* [cumulative product].

> *nrow(mtcars)*

*[1] 32*

> *mtcars$cyl*

*[1] 6 6 4 6 8 6 8 4 4 6 6 8 8 8 8 8 8 4 4 4 4 8 8 8 8 4 4 4 8 6 8 4*

> *pmin(mtcars$cyl)*

*[1] 6 6 4 6 8 6 8 4 4 6 6 8 8 8 8 8 8 4 4 4 4 8 8 8 8 4 4 4 8 6 8 4*

> *pmax(mtcars$cyl)*

*[1] 6 6 4 6 8 6 8 4 4 6 6 8 8 8 8 8 8 4 4 4 4 8 8 8 8 4 4 4 8 6 8 4*

> *cummin(mtcars$cyl)*

*[1] 6 6 4 4 4 4 4 4 4 4 4 4 4 4 4 4 4 4 4 4 4 4 4 4 4 4 4 4 4 4 4 4*

> *cummax(mtcars$cyl)*

*[1] 6 6 6 6 8 8 8 8 8 8 8 8 8 8 8 8 8 8 8 8 8 8 8 8 8 8 8 8 8 8 8 8*

> *cumsum(mtcars$cyl)*

*[1]   6  12  16  22  30  36  44  48  52  58  64  72  80  88  96 104 112 116 120 124*
*[21] 128 136 144 152 160 164 168 172 180 186 194 198*

> *cumprod(mtcars$cyl)*

*[1] 6.000000e+00 3.600000e+01 1.440000e+02 8.640000e+02 6.912000e+03*
                                                        *4.147200e+04*

*[7] 3.317760e+05 1.327104e+06 5.308416e+06 3.185050e+07 1.911030e+08*
                                                        *1.528824e+09*

*[13] 1.223059e+10 9.784472e+10 7.827578e+11 6.262062e+12 5.009650e+13*
                                                        *2.003860e+14*

*[19] 8.015440e+14 3.206176e+15 1.282470e+16 1.025976e+17 8.207810e+17*

*6.566248e+18*

*[25] 5.252999e+19 2.101199e+20 8.404798e+20 3.361919e+21 2.689535e+22*

*1.613721e+23*

*[31] 1.290977e+24 5.163908e+24*

## 5.2. Summary Statistics

Thus the *summary()* function can be applied on the entire dataset to get all the statistical values of all the numeric fields.

> *summary(mtcars)*

```
     mpg          cyl          disp          hp          drat
Min.   :10.40  Min.   :4.000  Min.   : 71.1  Min.   : 52.0  Min.   :2.760
1st Qu.:15.43  1st Qu.:4.000  1st Qu.:120.8  1st Qu.: 96.5  1st Qu.:3.080
Median :19.20  Median :6.000  Median :196.3  Median :123.0  Median :3.695
Mean   :20.09  Mean   :6.188  Mean   :230.7  Mean   :146.7  Mean   :3.597
3rd Qu.:22.80  3rd Qu.:8.000  3rd Qu.:326.0  3rd Qu.:180.0  3rd Qu.:3.920
Max.   :33.90  Max.   :8.000  Max.   :472.0  Max.   :335.0  Max.   :4.930

     wt           qsec          vs           am          gear
Min.   :1.513  Min.   :14.50  Min.   :0.0000  Min.   :0.0000  Min.   :3.000
1st Qu.:2.581  1st Qu.:16.89  1st Qu.:0.0000  1st Qu.:0.0000  1st Qu.:3.000
Median :3.325  Median :17.71  Median :0.0000  Median :0.0000  Median :4.000
Mean   :3.217  Mean   :17.85  Mean   :0.4375  Mean   :0.4062  Mean   :3.688
3rd Qu.:3.610  3rd Qu.:18.90  3rd Qu.:1.0000  3rd Qu.:1.0000  3rd Qu.:4.000
Max.   :5.424  Max.   :22.90  Max.   :1.0000  Max.   :1.0000  Max.   :5.000

     carb
Min.   :1.000
1st Qu.:2.000
Median :2.000
```

*Mean   :2.812*

*3rd Qu.:4.000*

*Max.   :8.000*

## 5.3. Normal Distribution

In a random collection of data from independent sources, it is generally observed that the distribution of data is normal. Which means, on plotting a graph with the value of the variable in the horizontal axis and the count of the values in the vertical axis we get a bell shape curve. The centre of the curve represents the mean of the data set. In the graph, half of values lie to the left of the mean and the other half lie to the right of the graph. This is referred as normal distribution in statistics. R has four in-built functions to generate normal distribution. They are described below.

*dnorm(x, mean, sd)*

*pnorm(x, mean, sd)*

*qnorm(p, mean, sd)*

*rnorm(n, mean, sd)*

*x - vector of numbers*

*p - vector of probabilities*

*n - sample size*

*mean - mean (default value is 0)*

*sd - standard deviation (default value is 1)*

## 5.3.1. dnorm()

For a given mean and standard deviation, this function gives the height of the probability distribution. Below is an example in which the result of the *dnorm()* function is plotted in a graph in *Fig. 5.1*.

```
> x <- seq(-5,5, by = 0.05)
> y <- dnorm(x, mean = 1.5, sd = 0.5)
> plot(x, y)
```

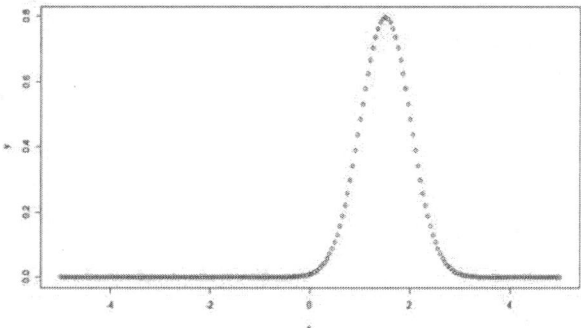

**Figure 5.1**  Plot of dnorm()

## 5.3.2. pnorm()

The *pnorm()* function returns the probability of a normally distributed random number which is less than the value of a given number. The other name for this is *"Cumulative Distribution Function"*. Below is an example in which the result of the *pnorm()* function is plotted in a graph as in *Fig. 5.2*.

> *x <- seq(-5,5, by = 0.05)*

> *y <- pnorm(x, mean = 1.5, sd = 1)*

> *plot(x, y)*

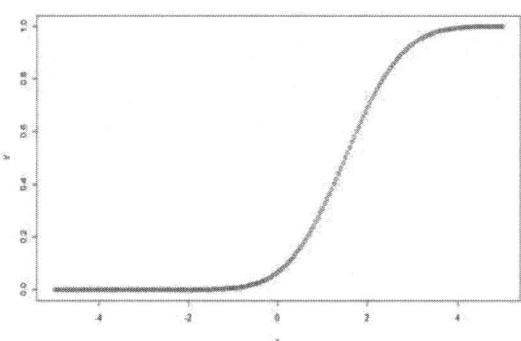

**Figure 5.2**  Plot of pnorm()

### 5.3.3. qnorm()

The *qnorm()* function takes the probability value as input and returns a cumulative value that matches the probability value. Below is an example in which the result of the *qnorm()* function is plotted in a graph as in *Fig. 5.3*.

> ```
> > x <- seq(0, 1, by = 0.02)
> > y <- qnorm(x, mean = 2, sd = 1)
> > plot(x, y)
> ```

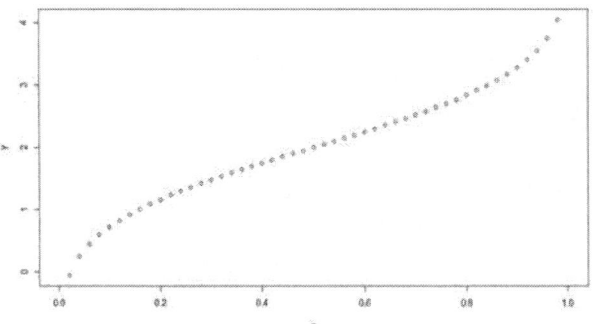

**Figure 5.3**  Plot of qnorm()

### 5.3.4. rnorm()

This function is used to generate random numbers whose distribution is normal. It takes the sample size as input and generates that many random numbers. We draw a histogram to show the distribution of the generated numbers as in *Fig. 5.4*.

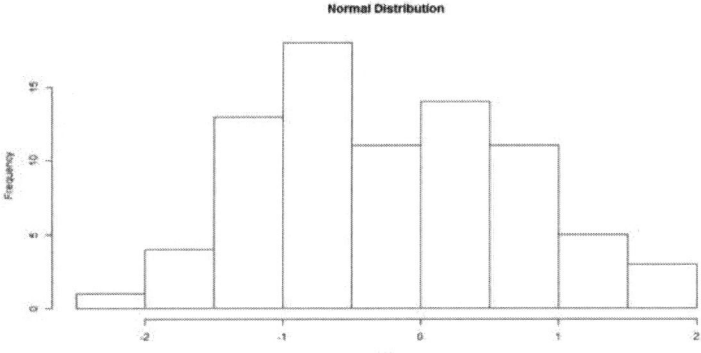

**Figure 5.4**   Histogram Using rnorm()

> *x <- rnorm(80)*

> *hist(x, main = "Normal Distribution")*

## 5.4. Binomial Distribution

The probability of success of an event is found by the binomial distribution model and this has only two possible outcomes in a series of experiments. For example, tossing of a coin always gives a head or a tail. During the binomial distribution, the probability of finding exactly 3 heads when tossing a coin for 10 times is estimated. R has four in-built functions to generate binomial distribution. They are described below.

*dbinom(x, size, prob)*

*pbinom(x, size, prob)*

*qbinom(p, size, prob)*

*rbinom(n, size, prob)*

*x - vector of numbers*

*p - vector of probabilities*

*n - sample size*

*size – number of trials*

*prob – probability of success of each trial*

## 5.4.1. dbinom()

This function gives the probability density distribution at each point. Below is an example in which the result of the *dbinom()* function is plotted in a graph as in *Fig. 5.5*.

> *x <- seq(0, 25, by = 1)*

> *y <- dbinom(x,25,0.5)*

> *plot(x, y)*

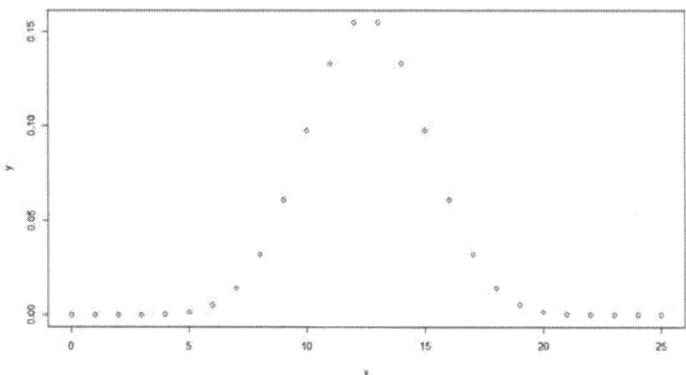

**Figure 5.5**   Plot Using dbinorm()

## 5.4.2. pbinom()

This function gives the cumulative probability of an event. It is a single value representing the probability. The probability of getting 25 or less heads from a 50 tosses of a coin is given by the below code.

> *x <- pbinom(25,50,0.5)*

> *x*

*[1] 0.5561376*

### 5.4.3. qbinom()

The function *qbinom()* takes the probability value as input and returns a number whose cumulative value matches the probability value. The below example finds how many heads will have a probability of 0.5 will come out when a coin is tossed 50 times.

```
> x <- qbinom(0.5, 50, 1/2)

> x

[1] 25
```

### 5.4.4. rbinom()

The function *rbinom()* returns the required number of random values of the given probability from a given sample. The below code is to find 5 random values from a sample of 50 with a probability of 0.5.

```
> x <- rbinom(5,50,0.5)

> x

[1] 24 21 22 29 32
```

## 5.5. Correlation Analysis

To evaluate the relation between two or more variables, the correlation test is used. Correlation coefficient in R can be computed using the functions *cor()* or *cor.test()*. The basic syntax for the correlation functions in R are as below.

*cor(x, y, method)*

*cor.test(x, y, method)*

*x, y - numeric vectors with the same length*

*method - correlation method ("pearson" or "kendall" or "spearman")*

Consider the data set *"mtcars"* available in the R environment. Let us first find the correlation between the horse power (*"hp"*) and the mileage per gallon (*"mpg"*) of the cars and then between the horse power (*"hp"*) and the cylinder displacement (*"disp"*) of the cars. From the test we find that the horse power (*"hp"*) and the mileage per gallon ("*mpg*") of the cars have *negative correlation* (-0.7761684) and the horse power ("*hp*") and the cylinder displacement ("*disp*") of the cars have positive correlation (0.7909486).

> cor(mtcars$hp, mtcars$mpg, method = "pearson")

[1] -0.7761684

> cor.test(mtcars$hp, mtcars$mpg, method = "pearson")

Pearson's product-moment correlation

data: mtcars$hp and mtcars$mpg

t = -6.7424, df = 30, p-value = 1.788e-07

alternative hypothesis: true correlation is not equal to 0

95 percent confidence interval:

-0.8852686 -0.5860994

sample estimates:

  cor

 -0.7761684

> cor(mtcars$hp, mtcars$disp, method = "pearson")

[1] 0.7909486

> cor.test(mtcars$hp, mtcars$disp, method = "pearson")

Pearson's product-moment correlation

data: mtcars$hp and mtcars$disp

t = 7.0801, df = 30, p-value = 7.143e-08

alternative hypothesis: true correlation is not equal to 0

95 percent confidence interval:

 0.6106794 0.8932775

sample estimates:

  cor

 0.7909486

The correlation results can also be viewed graphically as in Fig. 5.6. The corrplot() function can be used to analyze the correlation between the various columns of a dataset, say mtcars. After this, the correlation between individual columns can be compared by plotting it in separate graphs as in Fig. 5.7 and Fig. 5.8.

> *library(corrplot)*

> *M <- cor(mtcars)*

> *corrplot(M, method = "number")*

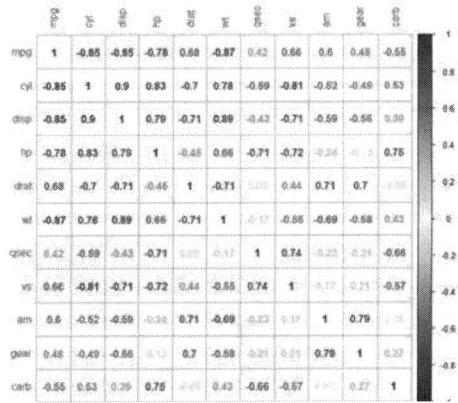

**Figure 5.6**   Cor Plot of mtcars Dataset

> *plot(mtcars$hp, mtcars$mpg, xlab="Horse Power of the Cars",*

> *ylab="Mileage per Gallon of the Cars", pch=21)*

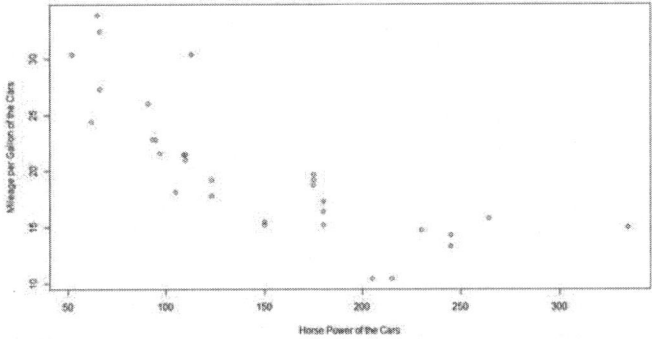

**Figure 5.7**   Scatter Plot of Negative Correlation

> *plot(mtcars$hp, mtcars$disp, xlab="Horse Power of the Cars",*

 *ylab="Cylinder Displacement of the Cars", pch=21)*

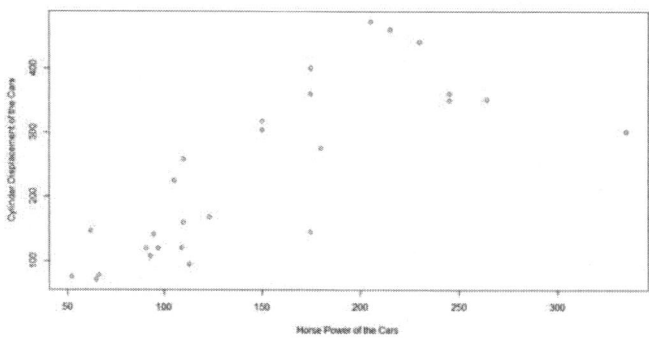

**Figure 5.8**   Scatter Plot of Positive Correlation

It can be noted that the graph with negative correlation (*Fig.* 5.7) has the dots from top left corner to bottom right corner and the graph with positive correlation (*Fig.* 5.8) has the dots from the bottom left corner to the top right corner.

# 5.6. Regression Analysis

## 5.6.1. Linear Regression

Regression analysis is a widely used statistical tool. A relationship model is established between the two variables used in the regression analysis. One of these variable is called *predictor* variable whose value is gathered through experiments. The other variable is called *response* variable whose value is derived from the predictor variable. Linear Regression of the two variables are related through an equation. The exponent of both the variables in this equation is 1. A linear relationship is represented by a *straight line* when plotted as a graph. A non-linear relationship where the exponent of any variable is not equal to 1 creates a curve.

The general mathematical equation for a linear regression is: $y = ax + b$, where $y$ is the response variable, $x$ is the predictor variable and $a$ and $b$ are constants which are called the coefficients.

A simple example of regression is predicting *income* of a person when his *expenditure* is known. To do this we need to have the relationship between income and expenditure of a person. First, carry out the experiment of gathering a sample of observed values of income and corresponding expenditures. Then create a relationship model using the *lm()* function in R. Find the coefficients from the model created and create the mathematical equation using these values. Now, get a summary of the relationship model to know the average error in prediction, which is also called *residuals*. Finally, to predict the income of the new persons, use the *predict()* function in R.

The function *lm()* creates the relationship model between the predictor and the response variable. The basic syntax for *lm()* function in linear regression is as given below.

*lm(formula,data)*

*formula - relation between x and y*

*data - numeric vector*

```
> x <- c(1510, 1740, 1380, 1860, 1280, 1360, 1790, 1630, 1520, 1310)
> y <- c(6300, 8100, 5600, 9100, 4700, 5700, 7600, 7200, 6200, 4800)
> model <- lm(y~x)
> model
```

*Call:*

*lm(formula = y ~ x)*

*Coefficients:*

| (Intercept) | x |
|---|---|
| -3845.509 | 6.746 |

```
> summary(model)
```

*Call:*

*lm(formula = y ~ x)*

*Residuals:*

| Min | 1Q | Median | 3Q | Max |
|---|---|---|---|---|
| -630.02 | -166.29 | 4.12 | 189.44 | 397.75 |

*Coefficients:*

*Estimate Std. Error t value Pr($> |t|$)*

*(Intercept) -3845.5087   804.9013  -4.778  0.00139 ***

*x            6.7461     0.5191  12.997 1.16e-06 ****

*---*

*Signif. codes:  0 '***' 0.001 '**' 0.01 '*' 0.05 '.' 0.1 ' ' 1*

*Residual standard error: 325.3 on 8 degrees of freedom*

*Multiple R-squared: 0.9548,        Adjusted R-squared: 0.9491*

*F-statistic: 168.9 on 1 and 8 DF,  p-value: 1.164e-06*

The basic syntax for the function *predict()* in linear regression is as given below.

*predict(object, newdata)*

*object - model created using lm() function*

*newdata - new numeric vector for predictor variable*

*> expense <- data.frame(x = 4700)*

*> income <- predict(model, expense)*

*> income*

*    1*

*27861.18*

The graphical representation of this linear regression is drawn by the below code and *Fig. 5.9.*

*> plot(y,x,col = "blue",main = "Income & Expenditure Regression",*

*abline(lm(x~y)),cex = 1.3,pch = 16,*

*xlab = "Income in Rs.",ylab = "Expenditure in Rs.")*

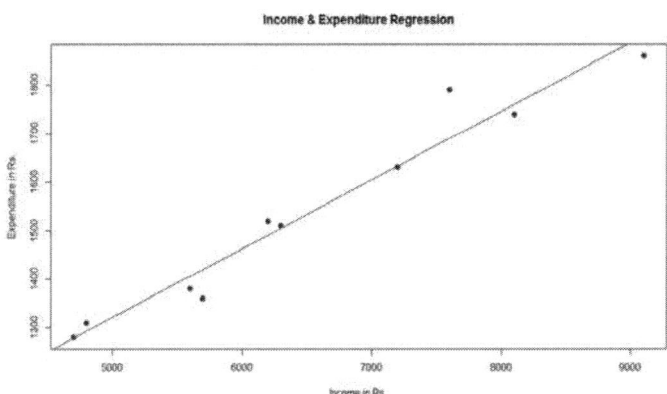

**Figure 5.9**  Plot of Linear Regression

## 5.6.2. Multiple Regressions

Multiple regressions is an extension of linear regression into relationship between more than two variables. In simple linear relation we have one predictor and one response variable, but in multiple regressions we have *more than one predictor* variable and *one response* variable.

The equation for multiple regressions consists of the below variables.

$$y = a + b_1 x_1 + b_2 x_2 + ... b_n x_n$$

In this equation $y$ is the response variable, $a$, $b_1$, $b_2$...$b_n$ are the coefficients and $x_1$, $x_2$, ...$x_n$ are the predictor variables.

In R, the *lm()* function is used to create the regression model. The model determines the value of the coefficients using the input data. Next we can predict the value of the response variable for a given set of predictor variables using these coefficients. The relationship model is built between the predictor variables and the response variables. The basic syntax for *lm()* function in multiple regression is as given below.

*lm(y ~ x1+x2+x3..., data)*

Consider the data set *"mtcars"* available in the R environment. This dataset presents the data of different car models in terms of mileage per gallon (*"mpg"*), cylinder displacement (*"disp"*), horse power (*"hp"*), weight of the car (*"wt"*) and some more parameters. This model establishes the relationship between *"mpg"* as a response variable with *"disp"*, *"hp"* and *"wt"* as predictor variables. We create a subset of these variables from the *mtcars* data set for this purpose.

> *model2 <- lm(mpg~disp+hp+wt, data = mtcars[, c("mpg", "disp", "hp", "wt")])*

> *model2*

*Call:*

*lm(formula = mpg ~ disp + hp + wt, data = mtcars[, c("mpg", "disp",*

   *"hp", "wt")])*

*Coefficients:*

*(Intercept)      disp      hp       wt*

  *37.105505   -0.000937   -0.031157   -3.800891*

> *a <- coef(model2)[1]*

> *a*

*(Intercept)*

  *37.10551*

> *b1 <- coef(model2)[2]*

> *b2 <- coef(model2)[3]*

> *b3 <- coef(model2)[4]*

> *b1*

    *disp*

*-0.0009370091*

> *b2*

    *hp*

*-0.03115655*

> *b3*

*wt*

-3.800891

We create the mathematical equation below, from the above intercept and coefficient values.

$$Y = a + b_1{}^*x_1 + b_2{}^*x_2 + b_3{}^*x_3$$
$$Y = (37.10551) + (-0.0009370091){}^*x_1 + (-0.03115655){}^*x_2 + (-3.800891){}^*x_3$$

We can use the regression equation created above to predict the mileage when a new set of values for displacement, horse power and weight is provided. For a car with disp = 160, hp = 110 and wt = 2.620 the predicted mileage is given by:

$$Y = (37.10551) + (-0.0009370091){}^*160 + (-0.03115655){}^*110 + (-3.800891){}^*2.620$$
$$= 23.57003$$

The above value can also be calculated using the function *predict()* for the given new value.

```
> newdata <- data.frame(disp = 160, hp = 110, wt = 2.620)
> mileage <- predict(model2, newdata)
> mileage
        1
23.57003
```

The graphical representation of this multiple regression is drawn by the below code and in *Fig. 5.10*.

```
> plot(mtcars$mpg, mtcars$disp+mtcars$hp+mtcars$wt, col = "blue",
    main = "Mileage Per Gallons & Other Factors Multiple Regression",
    abline(lm(mtcars$disp+mtcars$hp+mtcars$wt~mtcars$mpg)),
    cex = 1.5, pch = 15,
    xlab = "Other Facors (mtcars$disp+mtcars$hp+mtcars$wt)",
    ylab = "Mileage Per Gallon")
```

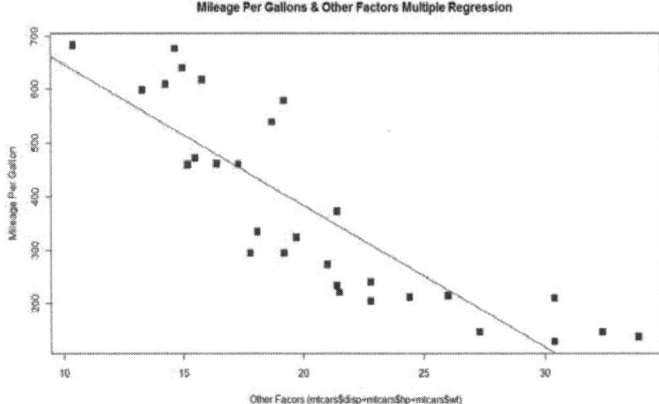

**Figure 5.10**   Plot of Multiple Regressions

### 5.6.3. Logistic Regression

The Logistic Regression is a regression model in which the response variable (dependent variable) has categorical values such as True/False or 0/1. It actually measures the probability of a binary response as the value of response variable based on the mathematical equation relating it with the predictor variables. The general mathematical equation for logistic regression is as given below.

$$y = 1 / ( 1 + e \char`\^ -(a + b_1x_1 + b_2x_2 + b_3x_3 + ... + b_nx_n))$$

In this equation y is the response variable, a, $b_1$, $b_2$...$b_n$ are the coefficients and $x_1$, $x_2$, ...$x_n$ are the predictor variables.

The function used to create the logical regression model is the *glm()* function. The basic syntax for *glm()* function in logistic regression is as given below.

*glm(formula,data,family)*

*formula - relation between the variables $x_1$, $x_2$, ...$x_n$ and y*

*data - numeric vectors of data*

*family - takes the value "binomial" for logistic regression*

The in-built data set *"mtcars"* describes different models of a car with their various engine specifications. In *"mtcars"* data set, the column *am* describes the transmission mode with a binary value. A logistic regression model is built between the columns *"am"* and 3 other columns - *hp, wt* and *cyl*.

> model3 <- glm(am ~ cyl + hp + wt, data = mtcars[, c("am", "cyl", "hp", "wt")],

family = binomial)

> model3

Call: glm(formula = am ~ cyl + hp + wt, family = binomial, data = mtcars[
    c("am", "cyl", "hp", "wt")])

Coefficients:

(Intercept)      cyl        hp        wt
 19.70288    0.48760    0.03259   -9.14947

Degrees of Freedom: 31 Total (i.e. Null);  28 Residual

Null Deviance:        43.23

Residual Deviance: 9.841   AIC: 17.84

> summary(model3)

Call:

glm(formula = am ~ cyl + hp + wt, family = binomial, data = mtcars[,
                                    c("am", "cyl", "hp", "wt")])

Deviance Residuals:

   Min     1Q    Median     3Q      Max
-2.17272 -0.14907 -0.01464  0.14116  1.27641

Coefficients:

     Estimate Std. Error z value Pr(> |z|)
(Intercept) 19.70288   8.11637   2.428   0.0152 *

| | | | | |
|---|---|---|---|---|
| *cyl* | 0.48760 | 1.07162 | 0.455 | 0.6491 |
| *hp* | 0.03259 | 0.01886 | 1.728 | 0.0840 . |
| *wt* | -9.14947 | 4.15332 | -2.203 | 0.0276 * |

---

*Signif. codes:* 0 '***' 0.001 '**' 0.01 '*' 0.05 '.' 0.1 ' ' 1

*(Dispersion parameter for binomial family taken to be 1)*

   Null deviance: 43.2297 on 31 degrees of freedom

Residual deviance: 9.8415 on 28 degrees of freedom

AIC: 17.841

*Number of Fisher Scoring iterations: 8*

The *p-value* in the summary is greater than 0.05 for the variables "*cyl*" (0.6491) and "*hp*" (0.0840). This value is considered to be insignificant in contributing to the value of the variable "*am*". Only weight "*wt*" (0.0276) impacts the "*am*" value in this regression model.

## 5.6.4. Poisson Regression

Poisson Regression involves regression models in which the response variable is in the form of counts and not fractional numbers. For example, the count of number of births or number of wins in a football match series. Also the values of the response variables follow a Poisson distribution. The Poisson Regression is represented by the below equation.

$$log(y) = a + b_1 x_1 + b_2 x_2 + b_n x_n .....$$

In this equation $y$ is the response variable, $a$, $b_1$, $b_2$...$b_n$ are the coefficients and $x_1$, $x_2$, ...$x_n$ are the predictor variables.

The function used to create the Poisson regression model is the *glm()* function. The basic syntax for *glm()* function in logistic regression is as given below.

*glm(formula, data, family)*

*formula - relation between the variables $x_1$, $x_2$, ...$x_n$ and y*

*data - numeric vectors of data*

*family - takes the value "poisson" for poisson regression*

The data set *"warpbreaks"* describes the effect of wool type and tension on the number of warp breaks per loom. Let's consider *"breaks"* as the response variable which is a count of number of breaks. The wool *"type"* and *"tension"* are taken as predictor variables. The model so built shows the below results.

> *model4 <- glm(formula = breaks ~ wool + tension, data = warpbreaks,*

*family = poisson)*

> *summary(model4)*

*Call:*

*glm(formula = breaks ~ wool + tension, family = poisson, data = warpbreaks*

*Deviance Residuals:*

| Min | 1Q | Median | 3Q | Max |
|---|---|---|---|---|
| -3.6871 | -1.6503 | -0.4269 | 1.1902 | 4.2616 |

*Coefficients:*

*Estimate Std. Error z value Pr(>|z|)*

| | Estimate | Std. Error | z value | Pr(>\|z\|) | |
|---|---|---|---|---|---|
| *(Intercept)* | 3.69196 | 0.04541 | 81.302 | < 2e-16 | *** |
| *woolB* | -0.20599 | 0.05157 | -3.994 | 6.49e-05 | *** |
| *tensionM* | -0.32132 | 0.06027 | -5.332 | 9.73e-08 | *** |
| *tensionH* | -0.51849 | 0.06396 | -8.107 | 5.21e-16 | *** |

---

*Signif. codes:  0 '***' 0.001 '**' 0.01 '*' 0.05 '.' 0.1 ' ' 1*

*(Dispersion parameter for poisson family taken to be 1)*

   *Null deviance: 297.37  on 53  degrees of freedom*

 *Residual deviance: 210.39  on 50  degrees of freedom*

  *AIC: 493.06*

*Number of Fisher Scoring iterations: 4*

In the summary we look for the p-value in the last column to be less than 0.05 to consider an impact of the predictor variable on the response variable. As seen the wooltype B having tension type M and H have impact on the count of breaks.

*p-value of woolB = 6.49e-05 = 0.0000649 < 0.05*

*p-value of tensionM = 9.73e-08 = 0.0000000973 < 0.05*

*p-value of tensionH = 5.21e-16 = 0.000000000000000521 < 0.05*

## 5.6.5. Non-Linear Least Square Regression

When modelling real world data for regression analysis, we observe that it is rarely the case that the equation of the model is a linear equation giving a linear graph. The equation of the model of real world data involves mathematical functions of higher degree. In such a scenario, the plot of the model gives a curve rather than a line. Both linear and non-linear regression aims to adjust the values of the model's parameters. This is to find the line or curve that comes nearer to the data. On finding these values we will be able to estimate the response variable with good accuracy.

In Least Square regression, a regression model is established. In this regression model, the sum of the squares of the vertical distances of different points from the regression curve is minimized. We generally start with a defined model and assume some values for the coefficients. The *nls()* function of R is then apllied to get the more accurate values and the confidence intervals. The basic syntax for creating a nonlinear least square test in R is as below.

*nls(formula, data, start)*

*formula - a nonlinear model formula including variables and parameters*

*data – a data frame*

*start - list or vector of starting estimates*

We will consider a nonlinear model with assumption of initial values of its coefficients. Next we will see what the confidence intervals of these assumed values so that we can judge how well these values fit into the model. Consider the below equation.

$$a = b_1{}^* x \wedge 2 + b_2$$

Let us assume the initial coefficients to be 1 and 3 and fit these values into *nls()* function.

```
> x <- c(1.6, 2.1, 2, 2.23, 3.71, 3.25, 3.4, 3.86, 1.19, 2.21)
> y <- c(5.19, 7.43, 6.94, 8.11, 18.75, 14.88, 16.06, 19.12, 3.21, 7.58)
> plot(x, y)
> model <- nls(y ~ b1*x^2+b2, start = list(b1 = 1,b2 = 3))
> new <- data.frame(x = seq(min(x), max(x), len = 100))
> lines(new$x, predict(model, newdata = new))
> res1 <- sum(resid(model) ^ 2)
> res1
[1] 1.081935
> res2 <- confint(model)
> res2
      2.5%    97.5%
b1 1.137708 1.253135
b2 1.497364 2.496484
```

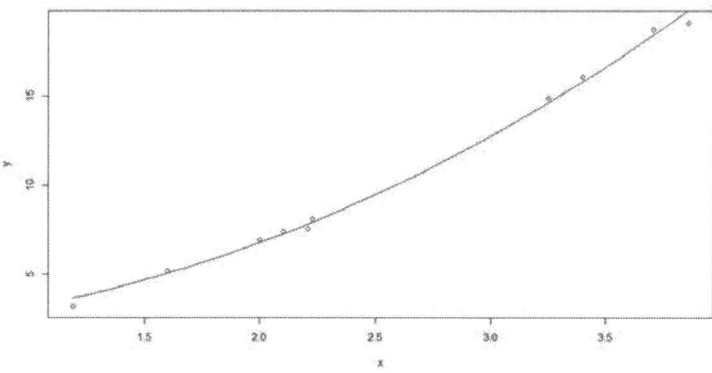

**Figure 5.11**   Plot of Non Linear Regressions

We can conclude that the value of $b_1$ is more close to 1 (*1.253135*) while the value of $b_2$ is more close to 2 (*2.496484*) and not 3.

# 5.7. Analysis of Variance (ANOVA)

Analysis of Variance (ANOVA) is a statistical measure that is used for investigating data by comparing the means of subsets of the data. The base case is the *one-way* ANOVA. In *one-way* ANOVA the data is sub-divided into groups based on a single classification factor. The one-way ANOVA is used to verify if the means of many groups are equals. But this analysis may not be very useful for complex problems. For example, it may be necessary to take into account two factors of variability to determine if the averages between the groups depend on the group classification or the second variable that is to consider. In this case the two-way analysis of variance (*two-way* ANOVA) should be used.

Consider the dataset *PlantGrowth* available in R for performing *one-way* ANOVA using R. This dataset has two columns, the control group / treatment and the weight of the plant indicating its growth. We want to check if the hypothesis that the control group / treatment has effect on the plant weight / plant growth. The below code does the same.

> *plant = lm(PlantGrowth$weight ~ PlantGrowth$group)*

> *anova(plant)*

*Analysis of Variance Table*

*Response: PlantGrowth$weight*

| | Df | Sum Sq | Mean Sq | F value | Pr(>F) |
|---|---|---|---|---|---|
| *PlantGrowth$group* | 2 | 3.7663 | 1.8832 | 4.8461 | 0.01591 * |
| *Residuals* | 27 | 10.4921 | 0.3886 | | |

*---*

*Signif. codes:  0 '***' 0.001 '**' 0.01 '*' 0.05 '.' 0.1 ' ' 1*

The result shows that the *F-value* is 4.8461 and the *p-value* is 0.01591 which is less than 0.05 (5% *level of significance*). This shows that the null hypothesis is rejected, that is the control group / treatment has effect on the plant growth / plant weight.

For *two-way* ANOVA, consider the below example of *revenues* collected for 5 *years* in *each month*. We want to see if the revenue depends on the Year and / or Month or if they are independent of these two factors.

```
> revenue = c(15,18,22,23,24, 22,25,15,15,14, 18,22,15,19,21,
+        23,15,14,17,18, 23,15,26,18,14, 12,15,11,10,8, 26,12,23,15,18,
+        19,17,15,20,10, 15,14,18,19,20, 14,18,10,12,23, 14,22,19,17,11,
+        21,23,11,18,14)

> months = gl(12,5)
> years = gl(5, 1, length(revenue))
> fit = aov(revenue ~ months + years)

> anova(fit)
Analysis of Variance Table

Response: revenue
```

|           | Df | Sum Sq | Mean Sq | F value | Pr(>F) |
|-----------|----|--------|---------|---------|--------|
| months    | 11 | 308.45 | 28.041  | 1.4998  | 0.1660 |
| years     | 4  | 44.17  | 11.042  | 0.5906  | 0.6712 |
| Residuals | 44 | 822.63 | 18.696  |         |        |

The significance of the difference between months is: $F = 1.4998$. This value is lower than the value tabulated and indeed *p-value* $> 0.05$. So we cannot reject the *null hypothesis*: the means of revenue evaluated according to the months are not proven to be not equal, hence we remain in our belief that the variable *"months" has no effect on revenue.*

The significance of the difference between years is: $F = 0.5906$. This value is lower than the value tabulated and indeed *p-value* $> 0.05$. So we fail to reject the *null hypothesis*: the means of revenue evaluated according to the years are not found to be un-equal, then the variable *"years" has no effect on revenue.*

# 5.8. Analysis of Covariance (ANCOVA)

We use Regression analysis to create models which describe the effect of variation in predictor variables on the response variable. Sometimes, if we have a categorical variable with values like Yes/No or Male/Female etc. the simple regression analysis gives multiple results for each value of the categorical variable. In this case, we can study the effect of categorical variable with the predictor variable. Such an analysis is termed as *Analysis of Covariance* also called as ANCOVA.

ANCOVA is a type of ANOVA model that has a general linear model with a continuous outcome variable and two or more predictor variables. Of these predictor variables, at least one is continuous and at least one more is categorical. *Analysis of variance* (ANOVA) is a collection of statistical models and their procedures which are used to observe differences between the means of three or more variables in a population based on the sample presented.

Consider the R built in data set *"mtcars"*. In this dataset the field *"am"* represents the type of transmission and it takes the values 0 or 1. The miles per gallon value, *"mpg"* of a car can also depend on it besides the value of horse power, *"hp"*. The effect of the value of *"am"* on the regression between *"mpg"* and *"hp"* is studied. It is done by using the *aov()* function followed by the *anova()* function to compare the multiple regressions.

Consider the fields *"mpg"*, *"hp"* and *"am"* from the data set *"mtcars"*. The variable *"mpg"* is the response variable, and the variable *"hp"* is chosen as the predictor variable and *"am"* as the categorical variable. We create a regression model taking *"hp"* as the predictor variable and *"mpg"* as the response variable taking into account the interaction between *"am"* and *"hp"*.

The model with interaction between categorical variable and predictor variable is given as below.

> res1 <- aov(mtcars$mpg ~ mtcars$hp * mtcars$am, data = mtcars)

> summary(res1)

    Df Sum Sq Mean Sq F value  Pr(>F)

---

| | | | | | |
|---|---|---|---|---|---|
| *mtcars$hp* | 1 | 678.4 | 678.4 | 77.391 | 1.50e-09 *** |
| *mtcars$am* | 1 | 202.2 | 202.2 | 23.072 | 4.75e-05 *** |
| *mtcars$hp:mtcars$am* | 1 | 0.0 | 0.0 | 0.001 | 0.981 |
| *Residuals* | 28 | 245.4 | 8.8 | | |

---

*Signif. codes:  0 '***' 0.001 '**' 0.01 '*' 0.05 '.' 0.1 ' ' 1*

As the *p-value* in both the cases is less than 0.05, the result shows that both horse power "*hp*" and transmission type "*am*" has significant effect on miles per gallon "*mpg*". But the interaction between these two variables is not significant as the *p-value* is more than 0.05.

The model without interaction between categorical variable and predictor variable is given as below.

> res2 <- aov(mtcars$mpg ~ mtcars$hp + mtcars$am, data = mtcars)

> summary(res2)

| | Df | Sum Sq | Mean Sq | F value | Pr(>F) |
|---|---|---|---|---|---|
| *mtcars$hp* | 1 | 678.4 | 678.4 | 80.15 | 7.63e-10 *** |
| *mtcars$am* | 1 | 202.2 | 202.2 | 23.89 | 3.46e-05 *** |
| *Residuals* | 29 | 245.4 | 8.5 | | |

---

*Signif. codes:  0 '***' 0.001 '**' 0.01 '*' 0.05 '.' 0.1 ' ' 1*

As the *p-value* in both cases is less than 0.05 in the above result, it shows that both horse power and transmission type has significant effect on miles per gallon. Now we can compare the two models to conclude if the interaction of the variables is truly in-significant. For this we use the *anova()* function.

> finres <- anova(res1,res2)

> finres

*Analysis of Variance Table*

*Model 1: mtcars$mpg ~ mtcars$hp * mtcars$am*

*Model 2: mtcars$mpg ~ mtcars$hp + mtcars$am*

| | Res.Df | RSS | Df | Sum of Sq | F | Pr(>F) |
|---|---|---|---|---|---|---|
| 1 | 28 | 245.43 | | | | |
| 2 | 29 | 245.44 | -1 | -0.0052515 | 6e-04 | 0.9806 |

As the *p-value* is (0.9806) greater than 0.05 we conclude that the interaction between horse power "*hp*" and transmission type "*am*" is not significant. So the mileage per gallon "*mpg*" will depend in a similar manner on the horse power of the car in both auto and manual transmission mode.

## 5.9. Chi Square Test

Chi-Square test determines if two categorical variables have a significant correlation between them. Both those variables should be from same population and they should be categorical like — Yes/No, Male/Female, Red/Green etc. For example, we can build a data set with observations on people's ice-cream buying pattern and try to correlate the gender of a person with the flavour of the ice-cream they prefer. If a correlation is found we can plan for appropriate stock of flavours by knowing the number of gender of people visiting.

The function *chisq.test()* is used for performing chi-Square test on the given data. The R syntax for chi-square test is as below.

*chisq.test(data)*

*data - data in form of a table*

We will take the *Cars93* data in the "MASS" library which represents the sales of different models of car in the year 1993. The facor variables in this dataset can be considered as categorical variables. In the below model the variables "*AirBags*" and "*Type*" are considered. Here we aim to find out any significant correlation between the types of car sold and the type of Air bags it has. If correlation is observed we can estimate which types of cars can sell better with what types of air bags.

```
> library(MASS)
> cardata = table(Cars93$AirBags, Cars93$Type)
> chi <- chisq.test(cardata)
> chi
    Pearson's Chi-squared test
data: cardata
X-squared = 33.001, df = 10, p-value = 0.0002722
```

The result shows the *p-value* (0.0002723) of less than 0.05 which indicates a strong correlation between the *"AirBags"* and *"Type"* of the cars sold.

## 5.10. Hypothesis Testing

The hypothesis is retained or rejected based on the *p-value* of the samples. This decision is based on the statistical concept of *hypothesis testing*. A *type I error* is the mishap of falsely *rejecting a null hypothesis when the null hypothesis is true*. The significance level of the hypothesis testing is the probability of committing a *type I error*. This is denoted by the Greek letter $\alpha$ (alpha). In hypothesis testing, a *type II error* is due to a *failure of rejecting an invalid null hypothesis*. The probability of avoiding a *type II error* is called the power of the hypothesis testing, and this is denoted by the quantity $1 - \beta$ *(1 − beta)*.

Hypothesis testing is of two types, *one-tailed tests* and *two-tailed tests*. A *one-tailed test* is a test of a hypothesis which is used to measure the relationship of variable in one direction which allows the hypothesis to be rejected. A *two-tailed test* is used to test the statistical significance of the hypothesis for the dataset to accept or reject the hypothesis.

### 5.10.1. Lower Tail Test

Suppose 60% of citizens voted in last election. 85 out of 148 people in a survey said that they voted in current election. At 0.5 significance level, is it possible to reject the null hypothesis that the proportion of voters in the population is above 60% this year? We apply the *prop.test()* function to compute the *p-value* directly. The *Yates continuity correction* is disabled.

> *prop.test(85, 148, p=.6, alt="less", correct=FALSE)*

*1-sample proportions test without continuity correction*

*data: 85 out of 148, null probability 0.6*
*X-squared = 0.40653, df = 1, p-value = 0.2619*
*alternative hypothesis: true p is less than 0.6*
*95 percent confidence interval:*
*0.0000000 0.6392527*
*sample estimates:*

*p*

0.5743243

The *p-value* is 0.5743 > 0.05. Hence, at .05 significance level, we do not reject the null hypothesis that the proportion of voters in the population is above 60% this year.

## 5.10.2. Upper Tail Test

Let 12% of apples taken from an orchard last year were rotten. This year, 30 out of 214 apples turns out to be rotten. Is it possible to reject the null hypothesis at .05 level of significance, that the proportion of rotten apples in harvest stays below 12% this year?

> *prop.test(30, 214, p=.12, alt="greater", correct=FALSE)*

*1-sample proportions test without continuity correction*

*data: 30 out of 214, null probability 0.12*

*X-squared = 0.82583, df = 1, p-value = 0.1817*

*alternative hypothesis: true p is greater than 0.12*

*95 percent confidence interval:*

*0.1056274 1.0000000*

*sample estimates:*

*p*

0.1401869

The *p-value* is *0.14018 > 0.05*. Hence, at .05 significance level, we do not reject the null hypothesis that the proportion of rotten apples in harvest stays below 12% this year.

## 5.10.3. Two-Tailed Test

Let 12 heads are turned up out of 20 trials in a coin toss. At .05 significance level, is it possible to reject the null hypothesis that the coin toss is fair?

> *prop.test(12, 20, p=0.5, correct=FALSE)*

*1-sample proportions test without continuity correction*

*data:  12 out of 20, null probability 0.5*

*X-squared = 0.8, df = 1, p-value = 0.3711*

*alternative hypothesis: true p is not equal to 0.5*

*95 percent confidence interval:*

*0.3865815 0.7811935*

*sample estimates:*

*p*

*0.6*

The *p-value* is *0.6 > 0.05*. Hence, at .05 significance level, we do not reject the null hypothesis that the coin toss is fair.

## ❖ HIGHLIGHTS

- Most important functions to get the statistical measures are available in the packages *base* and *stats*.
- The basic statistical measures are obtained by the functions *min()*, *max()*, *mean()* and *median()*.
- The basic statistical measures can also be obtained by one function *summary()*.
- R does not have a standard in-built function to calculate mode and hence we create a user function to calculate mode.

- The function *unique()* returns a vector, data frame or array with duplicate elements/rows removed.
- The function *match()* returns a vector of the positions of (first) matches of its first argument in its second.
- The function *tabulate()* takes the integer-valued vector bin and counts the number of times each integer occurs in it.
- The function *which.max()* determines the location, i.e., index of the (first) maximum of a numeric (or logical) vector.
- The functions to calculate the standard deviation, variance and the mean absolute deviation are *sd()*, *var()* and *mad()* respectively.
- The *quantile()* function provides the quartiles of the numeric values.
- The *IQR()* function provides the inter quartile range of the numeric fields.
- The function *cor()* and *cov()* are used to find the correlation and covariance between two numeric fields respectively.
- R has the built in functions to generate normal distribution of data namely *dnorm()*, *pnorm()*, *qnorm()* and *rnorm()*.
- R has the built in functions to generate binomila distribution of data namely *dbinom()*, *pbinom()*, *qbinom()* and *rbinom()*.
- Correlation coefficient in R can be computed using the functions cor() or *cor.test()*.
- The *corrplot()* function plots the correlation between the various columns of a dataset.
- The function *lm()* creates the relationship model between the predictor and the response variable in linear / multiple regression analysis.
- Multiple regressions is an extension of linear regression where we have more than one predictor variable and one response variable.
- The function used to create the logical / poisson regression model is the *glm()* function.
- The function used for Analysis of Variance (ANOVA) / Analysis of Covariance (ANCOVA) is *anova()* and *aov()*.
- The function used for performing chi-Square test is *chisq.test()*.
- In Hypothesis Testing, we apply the *prop.test()* function to compute the p-value directly.

# CHAPTER 6

# DATA MINING USING R

❖ **OBJECTIVES**

On completion of this Chapter you will be able to:

- know the main packages used for Data Mining in R
- know the packages and functions used for clustering in R
- apply k-means, k-medoids, hierarchical and density based clustering on datasets
- know the packages and functions used for classification in R
- apply packages party, rpart and randomForest for classifying datasets
- perform association rule mining on datasets and visualize the results
- perform univariate / multivariate outlier detection using R
- perform outlier detection using LOF and clustering techniques in R
- perform principal component analysis using R
- perform feature selection using R

## 6.1. Packages for Data Mining

Data Mining comprises of techniques and algorithms, for determining interesting patterns from large datasets. There are many algorithms that perform tasks such as frequent pattern mining, clustering, and classification. It is a continuous challenge faced by data mining analysts, to understand the importance of usage of these algorithms. There are a large number of implementations available in R for the various data mining algorithms currently present. Thus we can use the various packages and function available in R for performing Clustering, Classification,

Association Rule Mining, Outlier Detection and Dimensionality Reduction. The R packages that are related to data mining are *stats, cluster, fpc, sna, e1071, cba, biclust, clues, kohonen, rpart, party, randomForest, ada, caret, arules, eclat, arulesViz, DMwR, dprep, Rlof, plyr, corrplot, RWeka, gausspred, optimsimplex, CCMtools, FactoMineR* and *nnet*.

## 6.2. Clustering Using R

Clustering is a process of grouping a set of objects into clusters in such a way that objects in the same cluster are more similar to each other than to those in other clusters. The various clustering techniques are *Partitioning* Method, *Hierarchical* Method, *Density-based* Method, *Grid-Based* Method, *Model-Based* Method and *Constraint-based* Method. R Package provides various packages and functions that implement many clustering techniques such as *K-Means, Hierarchical, Fuzzy C-Means, PAM, SOM, CLARA, CLUES*, etc. R-Package has many clustering techniques implemented as the functions bundled in different library packages. We can see the below table listing the clustering techniques available in R along with their corresponding packages and functions.

### 6.2.1. Packages and Functions for Clustering

| Clustering Algorithm | Package | Function |
|---|---|---|
| K-Means Clustering | stats | kmeans() |
| Partitioning Around Medoids (PAM) | cluster | pam() |
| | fpc | pamk() |
| Hierarchical Clustering | cluster | agnes() |
| | stats | hclust() |
| Dissimilarity Matrix Calculation | cluster | daisy() |
| Density Based Clustering | fpc | dbscan() |
| K-Cores Clustering | sna | kcores() |

| Clustering Algorithm | Package | Function |
|---|---|---|
| Fuzzy C-Means Clustering | e1071 | cmeans() |
| RockCluster | cba | rockCluster() |
| Biclust | biclust | biclust() |
| CLUES | clues | clues() |
| Self-Organizing Maps (SOM) | kohonen | som() |
| Proximus | cba | proximus() |
| CLARA | cluster | clara() |

## 6.2.2. K-Means Clustering

*K Means Clustering* is an unsupervised learning algorithm. It tries to cluster data based on their similarity. Unsupervised learning means that there is no outcome to be predicted. The algorithm just tries to find patterns in the data. It is required to specify the number of clusters needed in the data to be grouped. This algorithm assigns randomly each observation to a cluster, and finds the cluster centroid. Then, the algorithm repeats the below two steps: 1) Assign data to the cluster whose centroid is closest. 2) For each cluster, calculates the new centroid. The above two steps are repeated until the difference between the clusters cannot be further reduced. The within cluster variation is calculated as the sum of the *Euclidean distance* between the data and the cluster *centroids*.

R has the default dataset called *iris* and this dataset will be used in the example below. The cluster number is set to 3 in the below clustering as the number of distinct species in the iris dataset is 3 ("*setosa*", "*versicolor*", "*virginica*"). For the purpose of initial manipulation let us copy the *iris* dataset into another dataset called *iris1*. Then we remove the column "*Species*" from the dataset *iris1* and then apply *kmeans* clustering on it.

The result of the clustering is then compared with the "*Species*" column of the dataset *iris* to see if similar objects are grouped together. The result of clustering shows that the species "*setosa*" can be clustered separately and the other two species have few overlapping of objects and hence can be clustered together. The functions *kmeans()*, *table()*, *plot()* and *point()* are used below for getting and plotting the results. It can be noted that the plots can be drawn with any two dimensions of the species available at a particular time (*eg. Sepal.Length* Vs. *Sepal.Width*). Also, the results of the *kmeans* clustering can vary from run to run due to the selection of cluster centres.

> *iris1 <- iris*

> *iris1$Species <- NULL*

> *km <- kmeans(iris1, 3)*

> *km*

*K-means clustering with 3 clusters of sizes 50, 38, 62*

*Cluster means:*

|   | Sepal.Length | Sepal.Width | Petal.Length | Petal.Width |
|---|---|---|---|---|
| 1 | 5.006000 | 3.428000 | 1.462000 | 0.246000 |
| 2 | 6.850000 | 3.073684 | 5.742105 | 2.071053 |
| 3 | 5.901613 | 2.748387 | 4.393548 | 1.433871 |

*Clustering vector:*

*[1] 1 1 1 1 1 1 1 1 1 1 1 1 1 1 1 1 1 1 1 1 1 1 1 1 1 1 1 1 1 1 1 1 1 1 1 1 1 1 1 1*

*[41] 1 1 1 1 1 1 1 1 1 1 3 3 2 3 3 3 3 3 3 3 3 3 3 3 3 3 3 3 3 3 3 3 3 3 3 3 3 3 2 3 3*

*[81] 3 3 3 3 3 3 3 3 3 3 3 3 3 3 3 3 3 3 3 3 3 2 3 2 2 2 2 3 2 2 2 2 2 3 3 2 2 2 2 3*

*[121] 2 3 2 3 2 2 3 3 2 2 2 2 3 2 2 2 2 3 2 2 2 3 2 2 2 3 2 2 3 2 2 3*

*Within cluster sum of squares by cluster:*

*[1] 15.15100 23.87947 39.82097*

 *(between_SS / total_SS =  88.4 %)*

*Available components:*

*[1] "cluster"    "centers"    "totss"    "withinss"    "tot.withinss"*

*[6] "betweenss"    "size"    "iter"    "ifault"*

> *table(iris$Species, km$cluster)*

|           | 1  | 2  | 3  |
|-----------|----|----|----|
| *setosa*     | 50 | 0  | 0  |
| *versicolor* | 0  | 2  | 48 |
| *virginica*  | 0  | 36 | 14 |

> *plot(iris1[c("Sepal.Length", "Sepal.Width")], col = km$cluster)*

> *points(km$centers[, c("Sepal.Length", "Sepal.Width")],*

> *col = 1:3, pch = 8, cex = 2)*

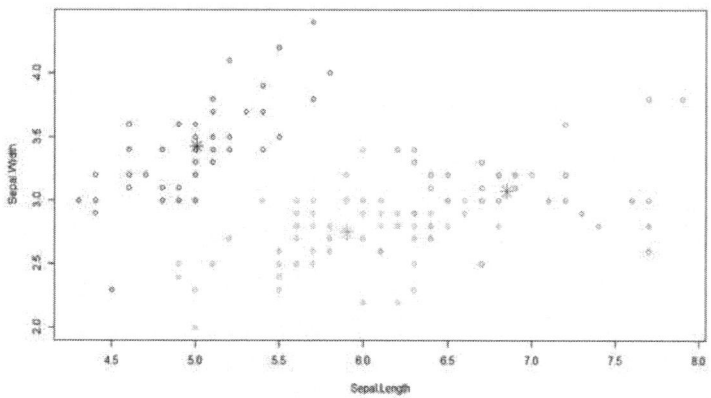

**Figure 6.1**　Plot of K-Means Clustering – Sepal.Length Vs. Sepal.Width

> *plot(iris1[c("Petal.Length", "Petal.Width")], col = km$cluster)*

> *points(km$centers[, c("Petal.Length", "Petal.Width")], col = 1:3, pch = 8, cex = 2)*

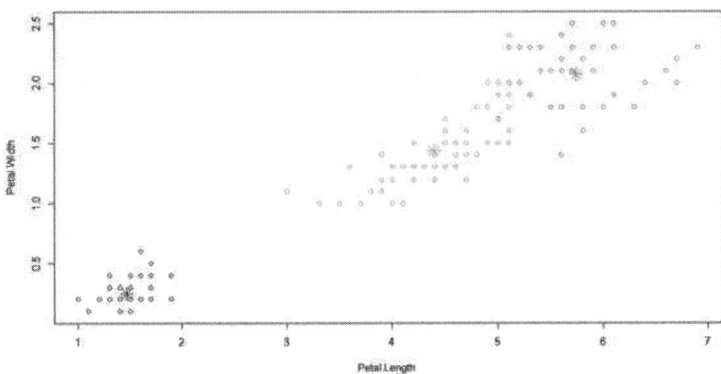

**Figure 6.2** Plot of K-Means Clustering – Petal.Length Vs. Petal.Width

## 6.2.3. K-Medoids Clustering

The *K-Medoids Clustering* algorithm is a partitioning clustering algorithm related to the *K-Means* algorithm. Both the *K-Means* and *K-Medoids* algorithms are partitional (breaking the dataset up into groups) and both attempt to minimize the distance between points labelled to be in a cluster and a point designated as the centre of that cluster. In contrast to the *K-Means* algorithm, *K-Medoids* chooses data points as centres and works with a generalization of the *Manhattan Norm* to define distance between data points. This method clusters the data set of *n* objects into *k* clusters. The *silhouette* is a useful tool for determining the number of clusters *k*.

R has the *pam()* and *pamk()* functions of the cluster package to do the *k-medoids* clustering. The *k-means* and *k-medoids* clustering produces almost the same result and the only difference is that in *k-means* the cluster is represented by the cluster centre and in *k-medoids* the cluster is represented by the object close to the cluster centre. But in the presence of outliers, *k-medoids* is more robust than *k-means* clustering. *Partitioning Around Medoids* (PAM) is the classic algorithm applied for *k-medoids* clustering. The PAM algorithm is not efficient in handling large datasets. *CLARA* is an enhanced technique of PAM which performs better on large datasets. For the functions *pam()* and *clara()* in the package *cluster* we need to specify the number of clusters. But, for the function *pamk()* in the package *fpc*, we need not specify the number of clusters. In this case the number of clusters is estimated using the *silhouette width*.

> *library(fpc)*

> *pmk <- pamk(iris1)*

> *pmk$nc*

*[1] 2*

> *table(pmk$pamobject$clustering, iris$Species)*

|   | setosa | versicolor | virginica |
|---|--------|------------|-----------|
| 1 | 50     | 1          | 0         |
| 2 | 0      | 49         | 50        |

> *layout(matrix(c(1, 2), 1, 2))*

> *plot(pmk$pamobject)*

> *layout(matrix(1))*

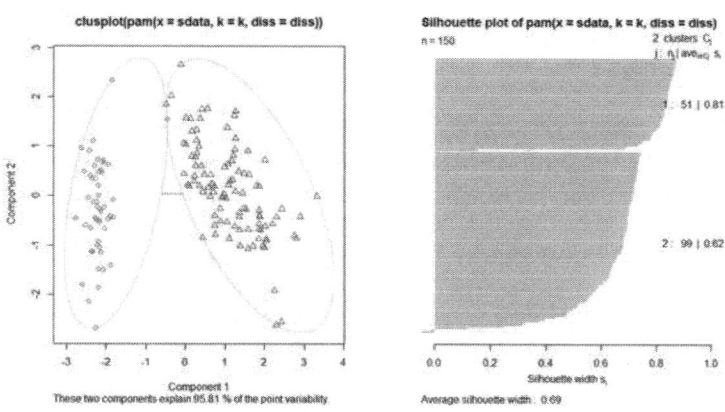

**Figure 6.3**   Plot of K-Medoids Clustering – Using pamk()

In the above left side chart of *Fig 4.3*, we can see that there are two clusters, one for the species "*setosa*" and the other for the mixture of species "*versicolor*" and "*virginica*". The right side chart of *Fig 4.3*, shows the *silhouette width* which decides the number of clusters (2 clusters in this case). The *silhouette width* is shown to be between 0.81 and 0.62 (*nearing 1*), and this means that the observations are well clustered. If the *silhouette width* is *around 0* it means the observations lies between the two clusters and it is *less than 0*, it means the observations are placed in the wrong clusters.

Now, let us use the *pam()* function from the *cluster* package to cluster the *iris* data and plot the results.

> *library(cluster)*

> *pm <- pam(iris1, 3)*

> *table(pm$clustering, iris$Species)*

|   | setosa | versicolor | virginica |
|---|--------|-----------|-----------|
| 1 | 50 | 0 | 0 |
| 2 | 0 | 48 | 14 |
| 3 | 0 | 2 | 36 |

> *layout(matrix(c(1, 2), 1, 2))*

> *plot(pm)*

> *layout(matrix(1))*

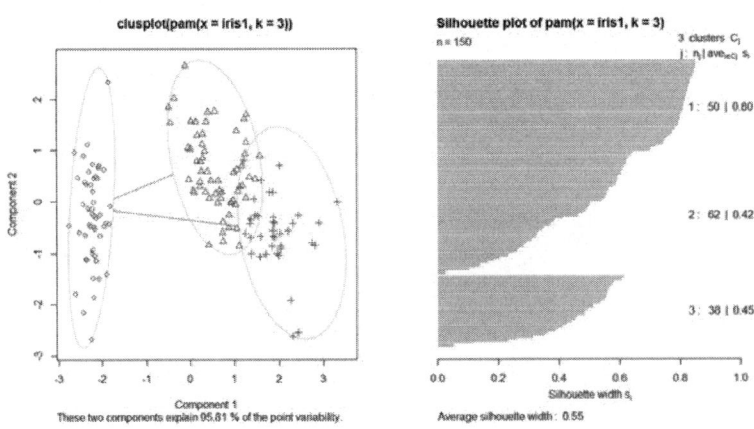

**Figure 6.4 – Plot of K-Medoids Clustering – Using pam()**

In the above left chart of *Fig. 6.4*, we can see three clusters. *Cluster 1* with only the *"setosa"* species and the *cluster 2* with mostly *"versicolor"* species and few of *"virginica"* species and in cluster 3 we have mostly *"viginica"* species and few of *"versicolor"* species. In both the above graphs *Fig. 6.3* and *Fig. 6.4*, the line between the clusters shows the distance between the clusters. From the above we can say that the choice of the clustering function used in R depends on the target problem and the domain knowledge available.

## 6.2.4. Hierarchical Clustering

*Hierarchical clustering* is a method of cluster analysis which seeks to build a hierarchy of clusters. Hierarchical clustering are of two types, Agglomerative and Divisive. *Agglomerative Hierarchical Clustering* is a *"bottom up"* approach. In this each observation starts with its own cluster, and pairs of clusters are merged into one cluster and this moves up the hierarchy. *Divisive Hierarchical Clustering* is a *"top down"* approach. In this all observations start in one cluster, and splits are performed recursively as one cluster moves down the hierarchy. The results of hierarchical clustering are usually presented in a *dendrogram*.

The hierarchical clustering in R can be done using the function *hclust()* in the *fpc* package. As the hierarchical clustering when plotted will be very crowded if the data is large, we create a sample of the *iris* data and do the clustering and plotting using this sample data. The function *rect.hclust()* is used to draw rectangle that covers each cluster. The *cutree()* function is used to draw the *dendrogram* as in the *Fig. 6.5*.

```
> i <- sample(1:dim(iris)[1], 50)
> iris3 <- iris[i,]
> iris3$Species <- NULL
> hc <- hclust(dist(iris3), method = "ave")
> plot(hc, hang = -1, labels = iris$Species[i])
> rect.hclust(hc, k=3)
> grp <- cutree(hc, k=3)
```

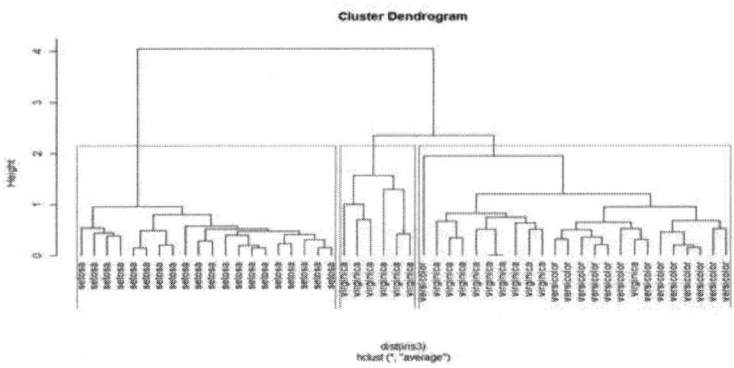

**Figure 6.5**   Plot of Hierarchical Clustering — Using hclust()

The resultant graph *Fig. 6.5*, shows that the first cluster has just the species "*setosa*", the second cluster has the species "*virginica*" and the third cluster has a mix of both the species "*versicolor*" and "*virginica*".

## 6.2.5. Density Based Clustering

*Density-based spatial clustering* of applications with noise (*DBSCAN*) is a most commonly used data clustering algorithm. In this clustering algorithm, given a set of points in some space, it groups together points that are closely packed together, marking as outliers points that lie alone in low-density regions.

The function *dbscan()* in the package *fpc* is used for the *density based* clustering. The *density based* clustering is done to cluster the entire data into one cluster. There are two main parameters in the function *dbscan()*. They are the *eps* and *MinPts*. The parameter *MinPts* stands for the minimum points in a cluster and the parameter *eps* defines the reachability distance. The *density based* clustering is sensitive to noisy data. Standard values are given for these parameters *eps* and *MinPts*.

> *library(fpc)*

> *iris4 <- iris*

> *iris4$Species <- NULL*

> *db <- dbscan(iris4, eps = 0.42, MinPts = 5)*

> *table(db$cluster, iris$Species)*

|   | setosa | versicolor | virginica |
|---|--------|------------|-----------|
| 0 | 2      | 10         | 17        |
| 1 | 48     | 0          | 0         |
| 2 | 0      | 37         | 0         |
| 3 | 0      | 3          | 33        |

> *plot(db, iris4)*

In the result above we can see that there are three clusters, *cluster 1*, *cluster 2* and *cluster 3*. The *cluster 0* corresponds to the outliers in the data.

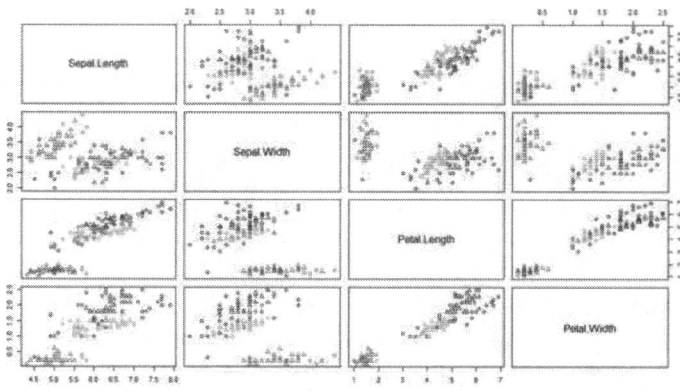

**Figure 6.6 – Plot of Density Based Clustering – Using dbscan()**

In the graph as in *Fig. 6.6*, the black circles represent the outliers. The clustering results can also be plotted in the scatter plot like in *k-means* clustering. This can be done using the function *plot()* as in *Fig. 6.7*, or the function *plotcluster()* as in *Fig. 6.8*, in the *fpc* package. The black circles in the *Fig. 6.7* and the black zeros "0" in the *Fig. 6.8* shows the outliers.

> *plot(db, iris4[c(1,2)])*

> *plotcluster(iris4, db$cluster)*

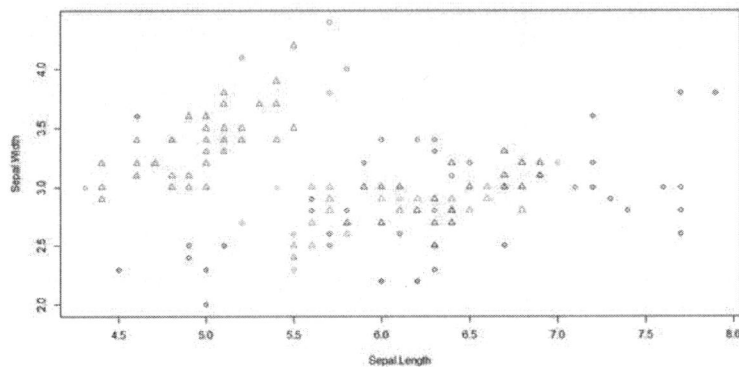

**Figure 6.7**   Scatter Plot of Density Based Clustering — Using plot()

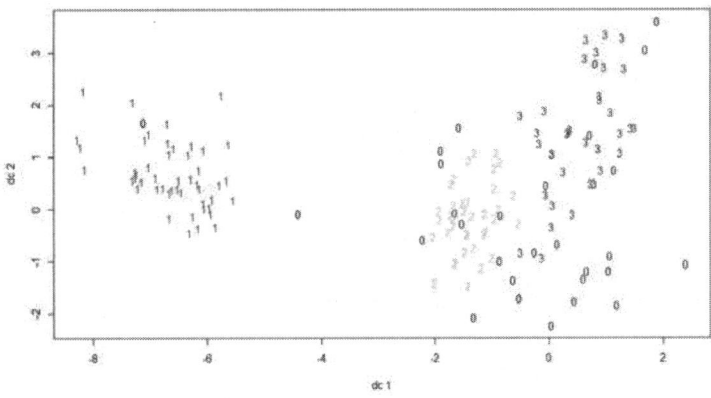

**Figure 6.8**   Scatter Plot of Density Based Clustering — Using plotcluster()

The clustering model can be used to model new data based on the similarity between the new data and the clusters. We take a sample of 10 records from the *iris* dataset and add some outliers to it and try to label the new dataset. The noises are generated using the function *runif()*.

> *set.seed(435)*

> *i <- sample(1:dim(iris)[1], 10)*

> *irissamp <- iris[1]*

> *irissamp <- iris[i, ]*

> *irissamp$Species <- NULL*

> *irissamp <- irissamp + matrix(runif(10\*4, min=0, max=0.3), nrow = 10, ncol = 4)*

> *plot(iris4[c(1,2)], col = db$cluster+1)*

> *points(irissamp[c(1,2)], pch=8, col=pred+1, cex = 2)*

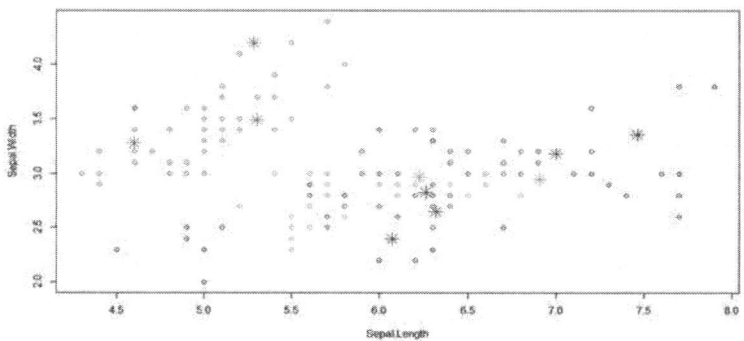

**Figure 6.9**   Scatter Plot of the Predicted Sample Clusters

> *pred <- predict(db, iris4, irissamp)*

> *table(pred, iris$Species[i])*

| pred | setosa | versicolor | virginica |
|------|--------|------------|-----------|
| 0 | 0 | 0 | 1 |
| 1 | 3 | 0 | 0 |
| 2 | 0 | 2 | 0 |
| 3 | 0 | 2 | 2 |

Thus from the above results and the plot in *Fig.* 6.9, we can see that out of the 10 new data, 7 (3 + 2 + 2) are assigned to the correct clusters and there is one outlier data in it.

# 6.3. Classification

Classification is a data mining technique that assigns items in a sample to target labelled classes. The goal of classification is to accurately predict the target class for each case in the data. For example, a classification model could be used to identify the age group of the people as children, youth, or old. The various classification techniques are *Decision Tree* based Methods, *Rule-based* Methods, *Memory based* reasoning, *Neural Networks*, *Naïve Bayes* and *Bayesian Belief Networks* and *Support Vector Machines*. R Package provides various packages and functions that implement many classification techniques such as *SVM*, *kNN*, *Decision Trees*, *Naive Bayes*, etc. R-Package has many classification techniques implemented as the functions bundled in different library packages. We can see the below table listing the classification techniques available in R along with their corresponding packages and functions.

## 6.3.1. Packages and Functions for Classification

| Clustering Algorithm | Package | Function |
|---|---|---|
| SVM | e1071 | svm() |
| kNN | RWeka | IBk() |
| Decision Trees | party | rpart() |
| | rpart | ctree()<br>cforest() |
| | randomForest | randomForest() |
| Naïve Bayes | e1071 | naiveBayes() |
| Adaboost | ada | ada() |
| JRip | caret | train() |

## 6.3.2. Decision Trees

Decision tree builds classification or regression models in the form of a tree structure. The decision tree is incrementally developed with the dataset being broken into smaller and smaller subsets. The final result is a tree with decision nodes and leaf nodes. A decision node has two or more branches. Leaf node represents a classification or decision. The topmost decision node in a tree is called the root node and this corresponds to the best predictor. Decision trees can handle both categorical and numerical data.

## 6.3.2.1. Package party

The function *ctree()* in the package *party* can be used to build the decision tree for the given data. We consider the *iris* dataset available in R for our analysis. This dataset has four attributes namely, the *Sepal.Length, Sepal.Width, Petal.Length* and *Petal.Width* using which we can predict the *Species* of the flower. After applying the function *ctree()* and getting the decision tree model, we can do the prediction using the function *predict()* for the given new data, so that we can categorize it into which *Species* the flowers belong to.

Before applying the decision tree function, the *iris* dataset is first split into training and test subsets. For training we choose 80% of the data randomly and the remaining 20% is used for testing. The seed for sampling is randomly set to a fixed number as below for effective splitting of data. After creating the decision tree using the training data, prediction is done on the test data. The results of the built tree can be viewed as text result and as well as a decision tree plot as in *Fig. 6.10*.

```
> set.seed(1234)
> i <- sample(2, nrow(iris), replace=TRUE, prob=c(0.8, 0.2))
> train <- iris[i==1,]
> test <- iris[i==2,]
> form <- Species ~ Sepal.Length + Sepal.Width +
                              Petal.Length + Petal.Width
> dt <- ctree(form, data=train)
> table(predict(dt), train$Species)
```

|            | setosa | versicolor | virginica |
|------------|--------|------------|-----------|
| setosa     | 42     | 0          | 0         |
| versicolor | 0      | 42         | 4         |
| virginica  | 0      | 1          | 34        |

> plot(dt)

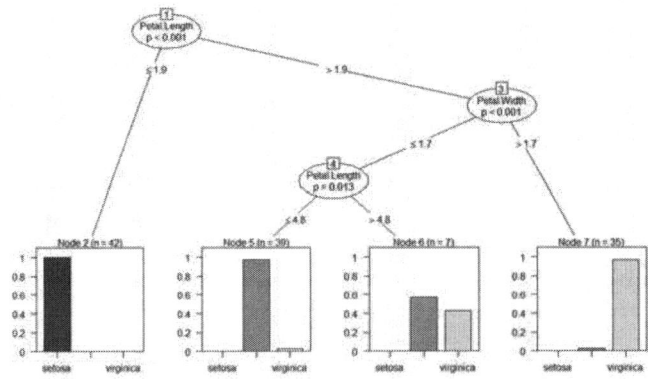

**Figure 6.10**    Decision Tree Plot — Using ctree()

> print(dt)

   *Conditional inference tree with 4 terminal nodes*

*Response: Species*

*Inputs: Sepal.Length, Sepal.Width, Petal.Length, Petal.Width*

*Number of observations: 123*

*1) Petal.Length <= 1.9; criterion = 1, statistic = 114.621*

  *2)\* weights = 42*

*1) Petal.Length > 1.9*

  *3) Petal.Width <= 1.7; criterion = 1, statistic = 54.728*

    *4) Petal.Length <= 4.8; criterion = 0.987, statistic = 8.717*

      *5)\* weights = 39*

    *4) Petal.Length > 4.8*

      *6)\* weights = 7*

3) *Petal.Width > 1.7*

7)\* *weights = 35*

The above decision tree result can also be drawn as a simple decision tree as in *Fig. 6.11*.

> *plot(dt, type = "simple")*

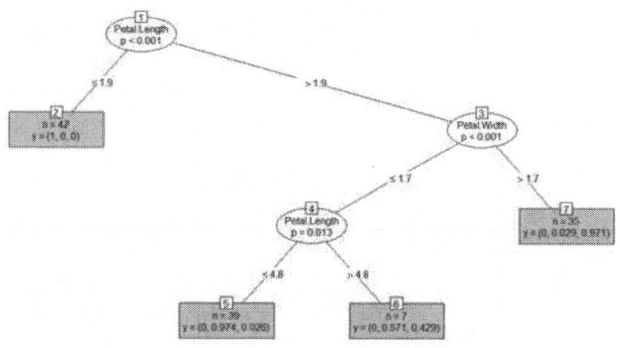

**Figure 6.11** Simple Decision Tree Plot – Using ctree()

In the first decision tree (*Fig. 6.10*) the number of training data under each species is listed as bar graph, but in the second decision tree (*Fig. 6.11*) the same is listed using variable *y*. For example, node 2 is labelled as "*n = 42, y(1, 0, 0)*", which means that it contains 42 training instances and all of them belong to the species "*setosa*".

Now, the predicted model will be tested with the test data to see if the instances are correctly classified.

> *pred <- predict(dt, newdata = test)*

> *table(pred, test$Species)*

| pred | setosa | versicolor | virginica |
|------|--------|------------|-----------|
| setosa | 8 | 0 | 0 |
| versicolor | 0 | 7 | 1 |
| virginica | 0 | 0 | 11 |

## 6.3.2.2. Package rpart

The function *rpart()* in the package *rpart* can be used to build the decision tree for the given data. We consider the *bodyfat* dataset available in the package *TH.data* of R for our analysis. After applying the function *rpart()* and getting the decision tree model, we can do the prediction using the function *predict()* for the given new data, so that we can categorize it into which *Species* the flowers belong to.

Before applying the decision tree function, the dataset is first split into training and test subsets. For training we choose 70% of the data randomly and the remaining 30% is used for testing. After creating the decision tree using the training data, prediction is done on the test data. The decision tree is shown in the *Fig. 6.12* and the details of the split are listed below.

```
> library(TH.data)
> data("bodyfat", package = "TH.data")
> set.seed(1234)
> i <- sample(2, nrow(bodyfat), replace=TRUE, prob=c(0.7, 0.3))
> train <- bodyfat[i==1,]
> test <- bodyfat[i==2,]
> form <- DEXfat ~ age + waistcirc + hipcirc +
                              elbowbreadth + kneebreadth
> dt2 <- rpart(form, data = train, control = rpart.control(minsplit = 10))
> plot(dt2)
> text(dt2, use.n=T, all = T, cex = 1)
```

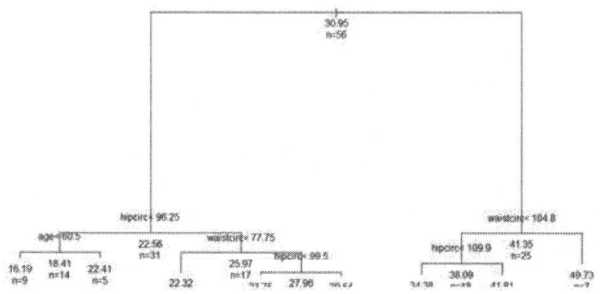

**Figure 6.12 – Classification Tree of bodyfat Dataset– Using rpart()**

> *str(bodyfat)*

*'data.frame':        71 obs. of  10 variables:*

*$ age       : num  57 65 59 58 60 61 56 60 58 62 ...*

*$ DEXfat     : num  41.7 43.3 35.4 22.8 36.4 ...*

*$ waistcirc  : num  100 99.5 96 72 89.5 83.5 81 89 80 79 ...*

*$ hipcirc    : num  112 116.5 108.5 96.5 100.5 ...*

*$ elbowbreadth: num  7.1 6.5 6.2 6.1 7.1 6.5 6.9 6.2 6.4 7 ...*

*$ kneebreadth : num  9.4 8.9 8.9 9.2 10 8.8 8.9 8.5 8.8 8.8 ...*

*$ anthro3a    : num  4.42 4.63 4.12 4.03 4.24 3.55 4.14 4.04 3.91 3.66 ...*

*$ anthro3b    : num  4.95 5.01 4.74 4.48 4.68 4.06 4.52 4.7 4.32 4.21 ...*

*$ anthro3c    : num  4.5 4.48 4.6 3.91 4.15 3.64 4.31 4.47 3.47 3.6 ...*

*$ anthro4     : num  6.13 6.37 5.82 5.66 5.91 5.14 5.69 5.7 5.49 5.25 ...*

> *print(dt2)*

*n= 56*

*node), split, n, deviance, yval*

*    * denotes terminal node*

*1) root 56 7265.0290000 30.94589*

*  2) waistcirc< 88.4 31  960.5381000 22.55645*

4) hipcirc< 96.25 14  222.2648000 18.41143

  8) age< 60.5 9  66.8809600 16.19222 *

  9) age>=60.5 5  31.2769200 22.40600 *

5) hipcirc>=96.25 17  299.6470000 25.97000

   10) waistcirc< 77.75 6  30.7345500 22.32500 *

   11) waistcirc>=77.75 11  145.7148000 27.95818

    22) hipcirc< 99.5 3  0.2568667 23.74667 *

    23) hipcirc>=99.5 8  72.2933500 29.53750 *

3) waistcirc>=88.4 25 1417.1140000 41.34880

  6) waistcirc< 104.75 18  330.5792000 38.09111

  12) hipcirc< 109.9 9  68.9996200 34.37556 *

  13) hipcirc>=109.9 9  13.0832000 41.80667 *

  7) waistcirc>=104.75 7  404.3004000 49.72571 *

```
> cp <- dt2$cptable[opt, "CP"]
> pred <- predict(dt2, newdata=test)
> xlim <- range(bodyfat$DEXfat)
> plot(pred ~ DEXfat, data=test, xlab="Observed",
                      ylab="Predicted", ylim=xlim, xlim=xlim)
> abline(a=0, b=1)
```

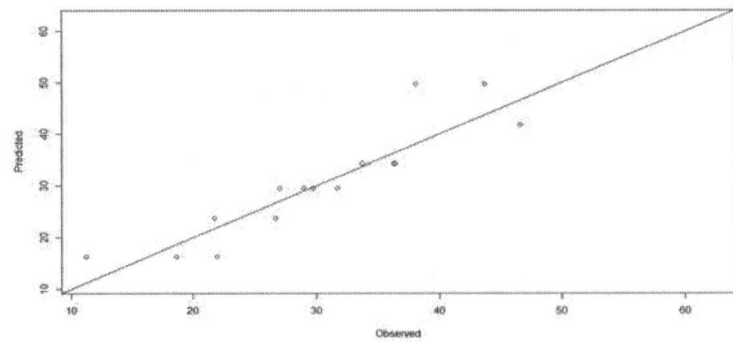

**Figure 6.13**  Graph of Predicted Vs. Observed Values

The predicted values are compared with the observed values and the graph in *Fig. 6.13* shows that the modelling is good as most points lie close to the diagonal line.

Similarly, we can also apply the same *rpart()* function for the *iris* dataset as before splitting the data into 80% training and 20% test data. The obtained model (*Fig. 6.14*) can then be used for predicting the species of the test data. The prediction shows that out of the 8 "*setosa*" species, all are correctly classified, out of the 8 "*versicolor*" species, 7 are correctly classified and 1 is incorrectly classified as "*virginica*" and out of the 11 "*virginica*" species, 10 are correctly classified and 1 is incorrectly classified as "*versicolor*".

```
> set.seed(1234)
> i <- sample(2, nrow(iris), replace=TRUE, prob=c(0.8, 0.2))
> train <- iris[i==1,]
> test < iris[i==2,]
> form <- Species ~ Sepal.Length + Sepal.Width +
                                    Petal.Length + Petal.Width
> dt <- rpart(form, data = train, control = rpart.control(minsplit = 10))
> table(predict(dt), train$Species)
```

|            | setosa | versicolor | virginica |
|------------|--------|------------|-----------|
| setosa     | 42     | 0          | 0         |
| versicolor | 0      | 42         | 4         |
| virginica  | 0      | 1          | 34        |

```
> plot(dt)
> text(dt, use.n=TRUE, all=TRUE)
```

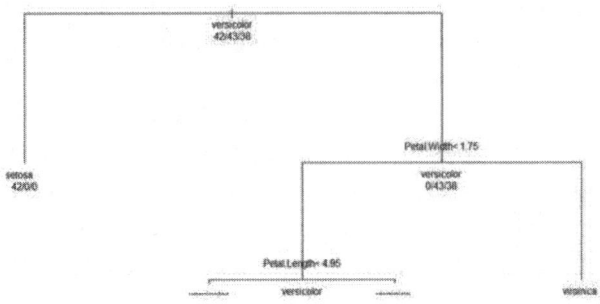

**Figure 6.14**   Classification Tree of Iris Dataset– Using rpart()

> print(dt)

n= 123

node), split, n, loss, yval, (yprob)

  * denotes terminal node

1) root 123 80 versicolor (0.34146341 0.34959350 0.30894309)

  2) Petal.Length< 2.45 42  0 setosa (1.00000000 0.00000000 0.00000000) *

  3) Petal.Length>=2.45 81 38 versicolor

                        0.00000000 0.53086420 0.46913580)

    6) Petal.Width< 1.75 46  4 versicolor

                    (0.00000000 0.91304348 0.08695652)

    12) Petal.Length< 4.95 41  1 versicolor

                    (0.00000000 0.97560976 0.02439024) *

    13) Petal.Length>=4.95 5  2 virginica

                    (0.00000000 0.40000000 0.60000000) *

  7) Petal.Width>=1.75 35  1 virginica

                    (0.00000000 0.02857143 0.97142857) *

> pred <- predict(dt, newdata=test)

> table(pred, test$Species)

| pred | setosa | versicolor | virginica |
|---|---|---|---|
| setosa | 8 | 0 | 0 |
| versicolor | 0 | 7 | 1 |
| virginica | 0 | 1 | 10 |

## 6.3.2.3. Package randomForest

The function *randomForest()* in the package *randomForest* can be used to classify the given data. We consider the same *iris* dataset available in R for our analysis. The *iris* dataset as before is split into 80% training and 20% test data. The obtained model (*Fig. 6.15*) can then be used for predicting the species of the test data.

```
> library(randomForest)
> set.seed(1234)
> i <- sample(2, nrow(iris), replace=TRUE, prob=c(0.8, 0.2))
> train <- iris[i==1,]
> test <- iris[i==2,]
> form <- Species ~ Sepal.Length + Sepal.Width +
                                    Petal.Length + Petal.Width
> rf <- randomForest(form, data=train, ntree=100, proximity=TRUE)
> plot(rf)
```

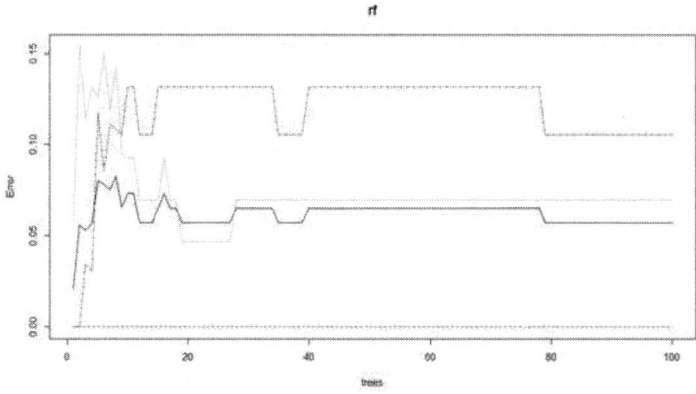

**Figure 6.15**   Random Forest of Iris Dataset

> *table(predict(rf), train$Species)*

|           | setosa | versicolor | virginica |
|-----------|--------|------------|-----------|
| setosa    | 42     | 0          | 0         |
| versicolor| 0      | 40         | 5         |
| virginica | 0      | 3          | 33        |

The importance of variables can be obtained with the functions *importance()* and *varImpPlot()* as in the *Fig. 6.16*.

> *importance(rf)*

|              | MeanDecreaseGini |
|--------------|------------------|
| Sepal.Length | 7.328201         |
| Sepal.Width  | 2.445821         |
| Petal.Length | 36.957590        |
| Petal.Width  | 34.436098        |

> *varImpPlot(rf)*

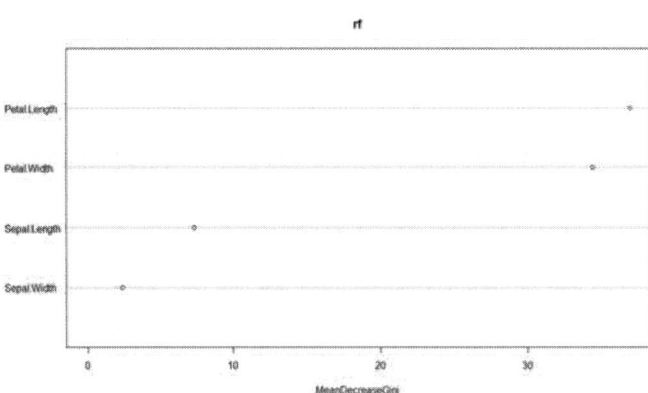

**Figure 6.16**   Importance of Variables in Iris Dataset

Finally, the built random forest is tested on test data, and the result is checked with the functions *table()* and *margin()*. The margin of a data point is the proportion of the correct class minus maximum proportion of the other classes. Generally, positive margin means correct classification (*Fig. 6.17*).

> *pred <- predict(rf, newdata=test)*

> *table(pred, test$Species)*

| pred | setosa | versicolor | virginica |
|------|--------|------------|-----------|
| setosa | 8 | 0 | 0 |
| versicolor | 0 | 7 | 0 |
| virginica | 0 | 0 | 12 |

> *plot(margin(rf, test$Species))*

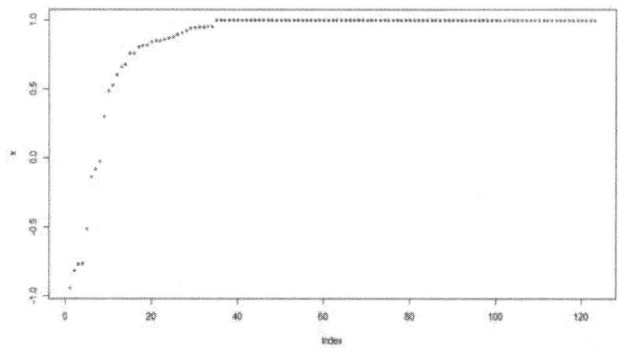

**Figure 6.17**   Margin of Classification by Random Forest Using Iris Dataset

Hence, by comparing the predictions of the above three functions *ctree()*, *rpart()* and *randomForest()*, we can see that the predictions of *ctree()* and *rpart()* are almost same and *randomForest()* prediction is little different as the *error rate (6.5%)* is high for the given data and sampling done

## 6.4. Association Rule Mining Using R

Association rules are rules presenting association or correlation between *item sets*. An association rule is in the form of A => B, where A and B are *two disjoint item sets*. A is called the *LHS* (left-hand side) and B is called the *RHS* (right-hand side) of the rule. The three most widely-used measures for selecting interesting rules are *support, confidence* and *lift. Support* is the percentage of cases in the data that contains both A and B. *Confidence* is the percentage of cases containing A that also contain B. *Lift* is the ratio of Confidence to the percentage of cases containing B. The formulae to calculate them are as below.

*support*(A => B) = P(A U B)

*confidence*(A => B) = P(B|A) = P(A U B) / P(A)

*lift*(A => B) = *confidence*(A => B) / P(B) = P(A U B) / P(A) P(B)

Where *P(A)* is the percentage (or probability) of cases containing A and *P(B)* is the percentage (or probability) of cases containing B.

The *Titanic* dataset in R is a 4-dimensional table with summarized information on the fate of passengers on the Titanic according to social class, sex, age and survival. We construct *titanic.raw* as raw data of the titanic dataset to make it suitable for applying association rule mining. In this *titanic.raw*, each row represents a person.

> *df <- as.data.frame(Titanic)*

> *titanic.raw <- NULL*

> *for(i in 1:4) { titanic.raw <- cbind(titanic.raw, rep(as.character(df[,i]), df$Freq)) }*

> *titanic.raw <- as.data.frame(titanic.raw)*

> *names(titanic.raw) <- names(df)[1:4]*

> *summary(titanic.raw)*

| Class | Sex | Age | Survived |
|---|---|---|---|
| 1st :325 | Female: 470 | Adult:2092 | No :1490 |
| 2nd :285 | Male :1731 | Child: 109 | Yes: 711 |
| 3rd :706 | | | |
| Crew:885 | | | |

A classic algorithm for association rule mining is APRIORI. It is a level-wise, breadth-first algorithm which counts transactions to find frequent item sets and then derive association rules from them. An implementation of it is function apriori() in package arules. Another algorithm for association rule mining is the ECLAT algorithm, which finds frequent item sets with equivalence classes, depth-first search and set intersection instead of counting. It is implemented using the function eclat() in the package eclat.

In the below section, we do association rule mining using the function *apriori()*. This function has the default settings as *supp=0.1*, which is the minimum support of rules, *conf=0.8*, which is the minimum confidence of rules and *maxlen=10*, which is the maximum length of the rules.

> *library(arules)*

> *rules.all <- apriori(titanic.raw)*

> *quality(rules.all) <- round(quality(rules.all), digits=3)*

> *rules.all*

   *set of 27 rules*

> *inspect(rules.all)*

| lhs | | rhs | support | confidence | lift |
|-----|---|-----|---------|------------|------|
| [1] {} | => | {Age=Adult} | 0.950 | 0.950 | 1.000 |
| [2] {Class=2nd} | => | {Age=Adult} | 0.119 | 0.916 | 0.964 |
| [3] {Class=1st} | => | {Age=Adult} | 0.145 | 0.982 | 1.033 |
| [4] {Sex=Female} | => | {Age=Adult} | 0.193 | 0.904 | 0.951 |
| [5] {Class=3rd} | => | {Age=Adult} | 0.285 | 0.888 | 0.934 |
| [6] {Survived=Yes} | => | {Age=Adult} | 0.297 | 0.920 | 0.968 |
| [7] {Class=Crew} | => | {Sex=Male} | 0.392 | 0.974 | 1.238 |
| [8] {Class=Crew} | => | {Age=Adult} | 0.402 | 1.000 | 1.052 |
| [9] {Survived=No} | => | {Sex=Male} | 0.620 | 0.915 | 1.164 |
| [10] {Survived=No} | => | {Age=Adult} | 0.653 | 0.965 | 1.015 |
| [11] {Sex=Male} | => | {Age=Adult} | 0.757 | 0.963 | 1.013 |
| [12] {Sex=Female,Survived=Yes} | => | {Age=Adult} | 0.144 | 0.919 | 0.966 |
| [13] {Class=3rd,Sex=Male} | => | {Survived=No} | 0.192 | 0.827 | 1.222 |
| [14] {Class=3rd,Survived=No} | => | {Age=Adult} | 0.216 | 0.902 | 0.948 |

*[15] {Class=3rd,Sex=Male}*

|  |  | => | *{Age=Adult}* | *0.210* | *0.906* | *0.953* |

*[16] {Sex=Male,Survived=Yes}*

|  |  | => | *{Age=Adult}* | *0.154* | *0.921* | *0.969* |

*[17] {Class=Crew,Survived=No}*

|  |  | => | *{Sex=Male}* | *0.304* | *0.996* | *1.266* |

*[18] {Class=Crew,Survived=No}*

|  |  | => | *{Age=Adult}* | *0.306* | *1.000* | *1.052* |

*[19] {Class=Crew,Sex=Male}*

|  |  | => | *{Age=Adult}* | *0.392* | *1.000* | *1.052* |

*[20] {Class=Crew,Age=Adult}*

|  |  | => | *{Sex=Male}* | *0.392* | *0.974* | *1.238* |

*[21] {Sex=Male,Survived=No}*

|  |  | => | *{Age=Adult}* | *0.604* | *0.974* | *1.025* |

*[22] {Age=Adult,Survived=No}*

|  |  | => | *{Sex=Male}* | *0.604* | *0.924* | *1.175* |

*[23] {Class=3rd,Sex=Male,Survived=No}*

|  |  | => | *{Age=Adult}* | *0.176* | *0.917* | *0.965* |

*[24] {Class=3rd,Age=Adult,Survived=No}*

|  |  | => | *{Sex=Male}* | *0.176* | *0.813* | *1.034* |

*[25] {Class=3rd,Sex=Male,Age=Adult}*

|  |  | => | *{Survived=No}* | *0.176* | *0.838* | *1.237* |

*[26] {Class=Crew,Sex=Male,Survived=No}*

|  |  | => | *{Age=Adult}* | *0.304* | *1.000* | *1.052* |

*[27] {Class=Crew,Age=Adult,Survived=No}*

|  |  | => | *{Sex=Male}* | *0.304* | *0.996* | *1.266* |

Many rules generated above are uninteresting. If we are interested in only rules with rhs indicating survival, we set rhs=c("Survived=No", "Survived=Yes") in the appearance argument to make sure that only "Survived=No" and "Survived=Yes" will appear in the rhs of rules. All other items can appear in the lhs, as set with the default="lhs". In the above result rules.all, we can also see that the left-hand side (lhs) of the first rule is empty. To exclude such rules, we set minlen to 2 in the code below. The details of progress are suppressed with verbose=F. After applying association rule mining, the resultant rules are sorted by lift so that the high-lift rules appear first. In the below code, the minimum support is set to 0.005, so each rule is supported at least by 12 (=ceiling(0.005 * 2201)) cases, which is acceptable for a population of 2201.

```
> rules <- apriori(titanic.raw, control = list(verbose=F),
    parameter = list(minlen=2, supp=0.005, conf=0.8),
    appearance = list(rhs=c("Survived=No", "Survived=Yes"), default="lhs"))
> quality(rules) <- round(quality(rules), digits=3)
> rules.sorted <- sort(rules, by="lift")
> inspect(rules.sorted)
```

| | lhs | rhs | support | confidence | lift |
|---|---|---|---|---|---|
| [1] | {Class=2nd,Age=Child} => | {Survived=Yes} | 0.011 | 1.000 | 3.096 |
| [2] | {Class=2nd,Sex=Female,Age=Child} => | {Survived=Yes} | 0.006 | 1.000 | 3.096 |
| [3] | {Class=1st,Sex=Female} => | {Survived=Yes} | 0.064 | 0.972 | 3.010 |
| [4] | {Class=1st,Sex=Female,Age=Adult} => | {Survived=Yes} | 0.064 | 0.972 | 3.010 |
| [5] | {Class=2nd,Sex=Female} => | {Survived=Yes} | 0.042 | 0.877 | 2.716 |
| [6] | {Class=Crew,Sex=Female} => | {Survived=Yes} | 0.009 | 0.870 | 2.692 |

*[7] {Class=Crew,Sex=Female,Age=Adult}*

        *=>*       *{Survived=Yes} 0.009*     *0.870*     *2.692*

*[8] {Class=2nd,Sex=Female,Age=Adult}*

        *=>*       *{Survived=Yes} 0.036*     *0.860*     *2.663*

*[9] {Class=2nd,Sex=Male, Age=Adult}*

        *=>*       *{Survived=No} 0.070*     *0.917*     *1.354*

*[10] {Class=2nd,Sex=Male}*

        *=>*       *{Survived=No} 0.070*     *0.860*     *1.271*

*[11] {Class=3rd,Sex=Male,Age=Adult}*

        *=>*       *{Survived=No} 0.176*     *0.838*     *1.237*

*[12] {Class=3rd,Sex=Male}*

        *=>*       *{Survived=No} 0.192*     *0.827*     *1.222*

Some rules generated above provide little or no extra information when some other rules are in the result. For example, the rule 2 provides no extra knowledge in addition to rule 1, since rules 1 tells us that all 2nd-class children survived. A rule is considered to be redundant when it is a super rule of another rule and it has the same or a lower lift. Other redundant rules in the above result are rules 4, 7 and 8, compared with the rules 3, 6 and 5 respectively. We prune redundant rules with code below.

```
> subset.matrix <- is.subset(rules.sorted, rules.sorted)
> subset.matrix[lower.tri(subset.matrix, diag=T)] <- FALSE
> redundant <- colSums(subset.matrix, na.rm=T) >= 1
> which(redundant)
  {Class=2nd,Sex=Female,Age=Child,Survived=Yes}
                       2
  {Class=1st,Sex=Female,Age=Adult,Survived=Yes}
                       4
  {Class=Crew,Sex=Female,Age=Adult,Survived=Yes}
                       7
```

{*Class=2nd,Sex=Female,Age=Adult,Survived=Yes*}

8

> *rules.pruned <- rules.sorted[!redundant]*

> *inspect(rules.pruned)*

| | lhs | rhs | support | confidence | lift |
|---|---|---|---|---|---|
| [1] | {*Class=2nd,Age=Child*} | | | | |
| | => | {*Survived=Yes*} 0.011 | 1.000 | 3.096 | |
| [2] | {*Class=1st,Sex=Female*} | | | | |
| | => | {*Survived=Yes*} 0.064 | 0.972 | 3.010 | |
| [3] | {*Class=2nd,Sex=Female*} | | | | |
| | => | {*Survived=Yes*} 0.042 | 0.877 | 2.716 | |
| [4] | {*Class=Crew,Sex=Female*} | | | | |
| | => | {*Survived=Yes*} 0.009 | 0.870 | 2.692 | |
| [5] | {*Class=2nd,Sex=Male,Age=Adult*} | | | | |
| | => | {*Survived=No*} 0.070 | 0.917 | 1.354 | |
| [6] | {*Class=2nd,Sex=Male*} | | | | |
| | => | {*Survived=No*} 0.070 | 0.860 | 1.271 | |
| [7] | {*Class=3rd,Sex=Male,Age=Adult*} | | | | |
| | => | {*Survived=No*} 0.176 | 0.838 | 1.237 | |
| [8] | {*Class=3rd,Sex=Male*} | | | | |
| | => | {*Survived=No*} 0.192 | 0.827 | 1.222 | |

The above rules show that only 2nd class children have survived. But, this cannot be the case as we have setup a higher support and confidence levels in the previous case. To investigate the above issue, we run the code below to find rules whose *rhs* is "*Survived=Yes*" and *lhs* contains "*Class=1st*", "*Class=2nd*", "*Class=3rd*", "*Age=Child*" and "*Age=Adult*" only, and which contains no other items (*default="none"*). We use lower thresholds for both support and confidence than before to find all rules for children of different classes.

```
> rules <- apriori(titanic.raw, parameter = list(minlen=3, supp=0.002, conf=0.2),
    appearance = list(rhs=c("Survived=Yes"),
    lhs=c("Class=1st", "Class=2nd", "Class=3rd", "Age=Child", "Age=Adult"),
    default="none"), control = list(verbose=F))
> rules.sorted <- sort(rules, by="confidence")

> inspect(rules.sorted)
```

| lhs | rhs | support | confidence | lift |
|---|---|---|---|---|
| [1] {Class=2nd,Age=Child} => {Survived=Yes} | 0.010904134 | 1.0000000 | 3.0956399 |
| [2] {Class=1st,Age=Child} => {Survived=Yes} | 0.002726034 | 1.0000000 | 3.0956399 |
| [3] {Class=1st,Age=Adult} => {Survived=Yes} | 0.089504771 | 0.6175549 | 1.9117275 |
| [4] {Class=2nd,Age=Adult} =>{Survived=Yes} | 0.042707860 | 0.3601533 | 1.1149048 |
| [5] {Class=3rd,Age=Child} => {Survived=Yes} | 0.012267151 | 0.3417722 | 1.0580035 |
| [6] {Class=3rd,Age=Adult} => {Survived=Yes} | 0.068605179 | 0.2408293 | 0.7455209 |

In the above result, the first two rules show that children of the 1st class are of the same survival rate as children of the 2nd class and that all of them survived. The rule of 1st-class children didn't appear before, simply because of its support was below the threshold specified. *Rule 5* presents a sad fact that children of class 3 had a low survival rate of 34%, which is comparable with that of 2nd-class adults and much lower than 1st-class adults.

Now, let us see some ways of visualizing association rules such as Scatter Plot, Grouped Matrix, Graph and Parallel Coordinates Plot as in *Fig. 6.18*, *Fig. 6.19*, *Fig. 6.20* and *Fig. 6.21* respectively.

```
> library(arulesViz)
> plot(rules.all)
> plot(rules.all, method="grouped")
```

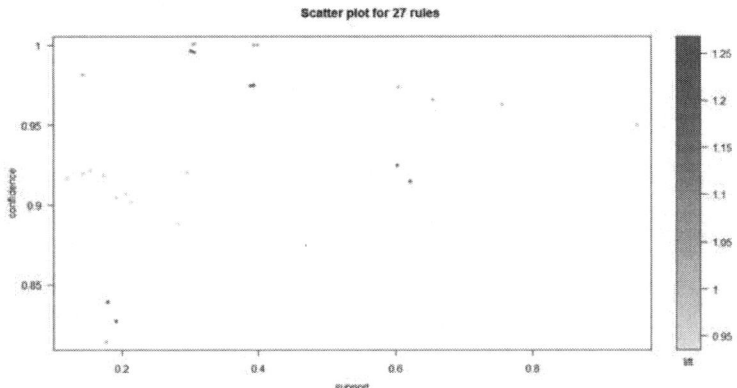

**Figure 6.18**  Scatter Plot of 27 Rules from Titanic Dataset

*> plot(rules.all, method="graph")*

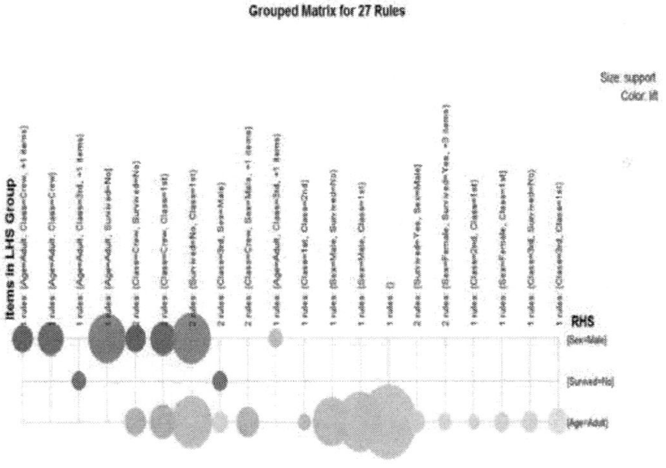

**Figure 6.19**  Grouped Matrix of 27 Rules from Titanic Dataset

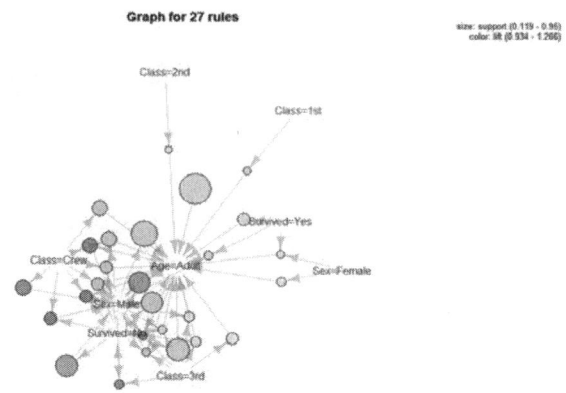

**Figure 6.20**   Graph of 27 Rules from Titanic Dataset

> *plot(rules.all, method="paracoord", control=list(reorder=TRUE))*

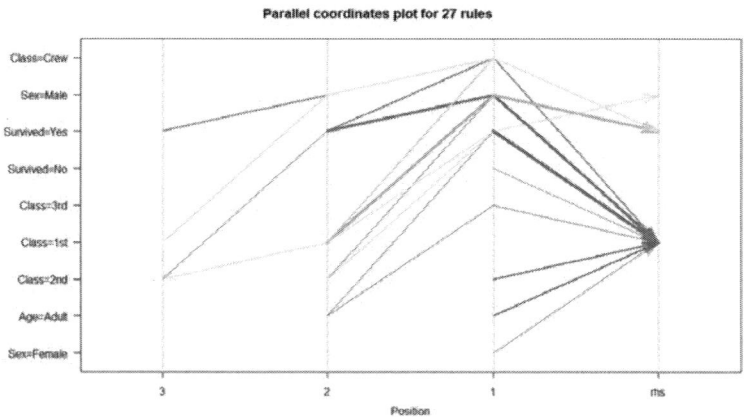

**Figure 6.21**   Parallel Coordinates Plot of 27 Rules from Titanic Dataset

# 6.5. Outlier Detection Using R

An outlier is a data point that is distant from other data points. An outlier may be caused due to difference in measuring or it may be due to error or oversight. There are two types of outliers, namely, univariate and multivariate outliers. A univariate outlier is an extreme data value of a particular variable.  A multivariate outlier is a combination of unusual data on minimum of two variables. These two types of outliers can influence the outcome of the data analysis. Outliers exist for four reasons. Incorrect data entry can cause data to contain extreme cases. Another reason for outliers can be missing values in a dataset. Yet, another possibility is that the data did not belong to the sample considered. And finally, the distribution of the sample for specific variables may have a more extreme distribution than normal.

## 6.5.1. Univariate Outlier Detection

Univariate outlier detection is done with function *boxplot.stats()*, which returns the statistics for producing boxplots. In the result returned by the function *boxplot. stats()*, one component is *out*, which gives a list of outliers. It lists the data points lying beyond the extremes of the *whiskers*. An argument of *coef* can be used to control how far the whiskers extend out from the box of a *boxplot*. The result of the same can be displayed as a box plot as in the *Fig. 6.22*.

```
> set.seed(3147)
> x <- rnorm(100)
> summary(x)
      Min. 1st Qu. Median   Mean 3rd Qu.   Max.
   -3.3154 -0.4837 0.1867 0.1098 0.7120 2.6859
> boxplot.stats(x)$out
[1] -3.315391  2.685922 -3.055717  2.571203
> boxplot(x)
```

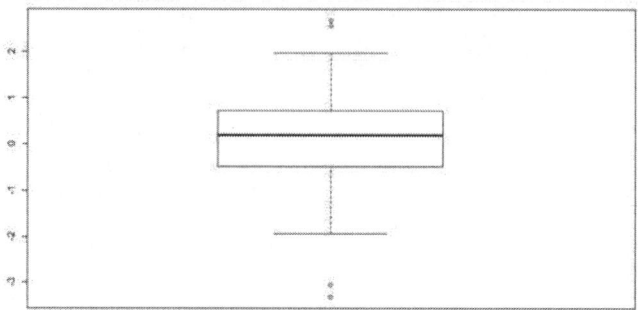

**Figure 6.22**   Box Plot of Univariate Outlier Detection

## 6.5.2. Multivariate Outlier Detection

The Univariate outlier detection can be used to find outliers in multivariate data also. First, we create a data frame with the two independent variables and detect their outliers separately. Then, we take multivariate outliers as those data which are outliers for both variables. In the below code outliers are marked with "+" in red. The result is displayed as a scatter plot as in *Fig. 6.23*.

> y <- rnorm(100)
> df <- data.frame(x, y)
> rm(x, y)
> head(df)

|   | x | y |
|---|---|---|
| 1 | -3.31539150 | 0.7619774 |
| 2 | -0.04765067 | -0.6404403 |
| 3 | 0.69720806 | 0.7645655 |
| 4 | 0.35979073 | 0.3131930 |
| 5 | 0.18644193 | 0.1709528 |
| 6 | 0.27493834 | -0.8441813 |

```
> a <- which(df$x %in% boxplot.stats(df$x)$out)
> b <- which(df$y %in% boxplot.stats(df$y)$out)
> a
[1]  1 33 64 74
> b
[1] 24 25 49 64 74
> outlierlist1 <- intersect(a,b)
> outlierlist1
[1] 64 74
> plot(df)
> points(df[outlierlist1,], col="red", pch="+", cex=2.5)
```

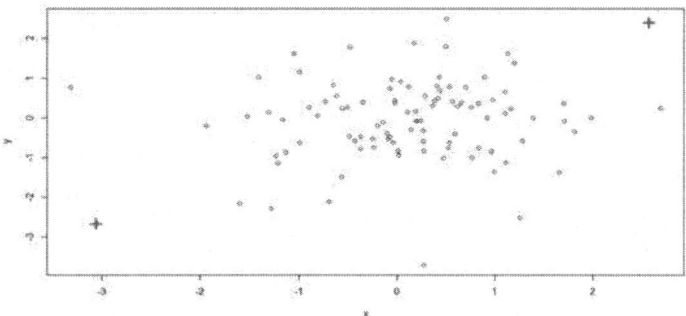

**Figure 6.23**   Scatter Plot of Multivariate Outlier Detection Using intersect()

Similarly, we can take multivariate outliers as those data which are outliers in either of the variables (x or y). This is shown as scatter plot in the *Fig. 6.24*.

```
> outlierlist2 <- union(a,b)
> outlierlist2
[1]  1 33 64 74 24 25 49
> plot(df)
> points(df[outlier.list2,], col="blue", pch="x", cex=2)
```

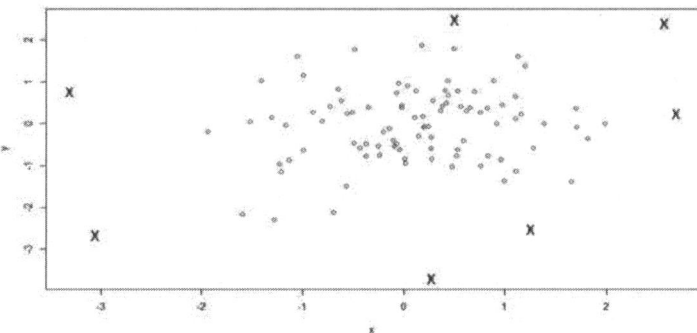

**Figure 6.24** Scatter Plot of Multivariate Outlier Detection Using union()

When there are three or more variables in an application, a final list of outliers might be produced with majority voting of outliers detected from individual variables. Domain knowledge should be involved when choosing the optimal way to group the items in real-world applications.

## 6.5.3. Outlier Detection Using LOF

*LOF* (Local Outlier Factor) is an algorithm for identifying density-based local outliers. With *LOF*, the local density of a point is compared with that of its neighbours. If the local density of this point is significantly lower than its neighbour (with an *LOF* value greater than one), the point is in a sparser region than its neighbours, which suggests it be an outlier. *LOF* works only on numeric data. Function *lofactor()* calculates local outlier factors using the *LOF* algorithm, and it is available in packages *DMwR* and *dprep*. An example of outlier detection with *LOF* is given below, where *k* is the number of neighbours used for calculating local outlier factors. A density plot of the outlier scores is also shown in the *Fig. 6.25*.

> *library(DMwR)*

> *iris1 <- iris[,1:4]*

> *outscores <- lofactor(iris1, k=5)*

> *plot(density(outscores))*

> *outliers <- order(outscores, decreasing=T)[1:5]*

> *print(outliers)*

*[1] 42 107 23 110 63*

> *print(iris1[outliers,])*

|  | Sepal.Length | Sepal.Width | Petal.Length | Petal.Width |
|---|---|---|---|---|
| 42 | 4.5 | 2.3 | 1.3 | 0.3 |
| 107 | 4.9 | 2.5 | 4.5 | 1.7 |
| 23 | 4.6 | 3.6 | 1.0 | 0.2 |
| 110 | 7.2 | 3.6 | 6.1 | 2.5 |
| 63 | 6.0 | 2.2 | 4.0 | 1.0 |

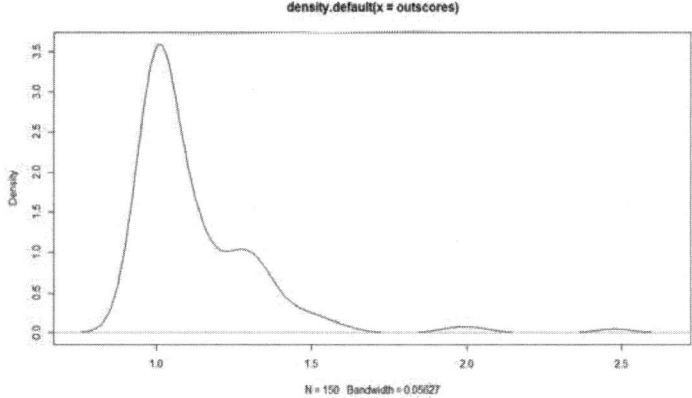

**Figure 6.25** Density Plot of Outliers Detection Using lofactor()

We show outliers with a biplot of the first two principal components in the *Fig. 6.26*. In the below code, *prcomp()* performs a principal component analysis, and *biplot()* plots the data with its first two principal components. In the below graph, *Fig. 6.26*, x-axis and y-axis are respectively the first and second principal components, the arrows show the original columns (variables), and the five outliers are labelled with their row numbers.

> *n <- nrow(iris1)*

> *labels <- 1:n*

> *labels[-outliers] <- "."*

> *biplot(prcomp(iris1), cex=.8, xlabs=labels)*

We can also show outliers with a *pairs plot* as below, where outliers are labelled with "+" in red.

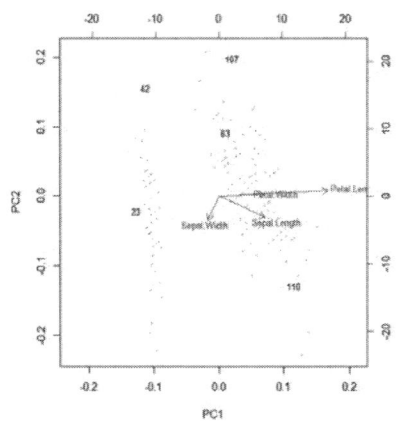

**Figure 6.26**    Biplot of First Two Principal Components

> *pch <- rep("\*", n)*

> *pch[outliers] <- "+"*

> *col <- rep("blue", n)*

> *col[outliers] <- "red"*

> *pairs(iris1, pch=pch, col=col, cex = 3)*

The outliers can also be displayed using the *pairs()* function as in the *Fig. 6.27* in which the outliers are marked as "+" symbol displayed in red colour. Package *Rlof* provides the function *lof()*, a parallel implementation of the *LOF* algorithm.

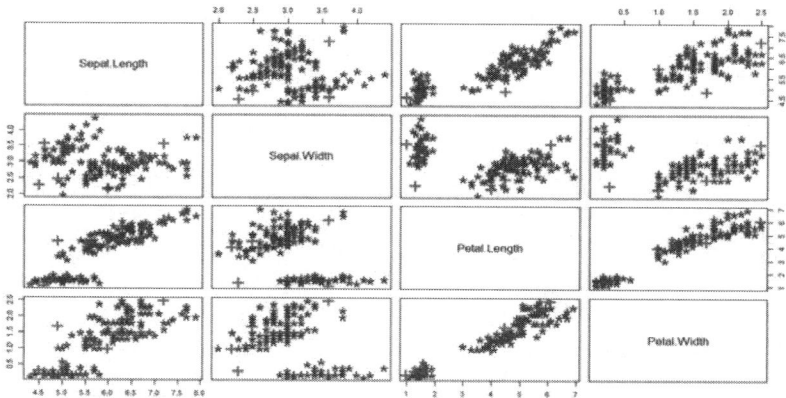

**Figure 6.27**   Outliers Displayed Using pairs()

## 6.5.4. Outlier Detection Using Clustering

One way to detect outliers is clustering. By grouping data into clusters, those data not assigned to any clusters are taken as outliers. For example, with density-based clustering such as *DBSCAN*, objects are grouped into one cluster if they are connected to one another by densely populated area. Therefore, objects not assigned to any clusters are isolated from other objects and are taken as outliers.

The function *dbscan()* in the package *fpc* is used for the *density based* clustering. There are two main parameters in the function *dbscan()*. They are the *eps* and *MinPts*. The parameter *MinPts* stands for the minimum points in a cluster and the parameter *eps* defines the reachability distance. Standard values are given for these parameters.

> *library(fpc)*

> *iris1 <- iris*

> *iris1$Species <- NULL*

> *db <- dbscan(iris1, eps = 0.42, MinPts = 5)*

> *table(db$cluster, iris$Species)*

|   | setosa | versicolor | virginica |
|---|--------|------------|-----------|
| 0 | 2 | 10 | 17 |
| 1 | 48 | 0 | 0 |
| 2 | 0 | 37 | 0 |
| 3 | 0 | 3 | 33 |

> plot(db, iris1)

In the result above we can see that there are three clusters, *cluster* 1, *cluster* 2 and *cluster* 3. The *cluster* 0 corresponds to the outliers in the data (marked by black circles in the plot of *Fig.* 6.28).

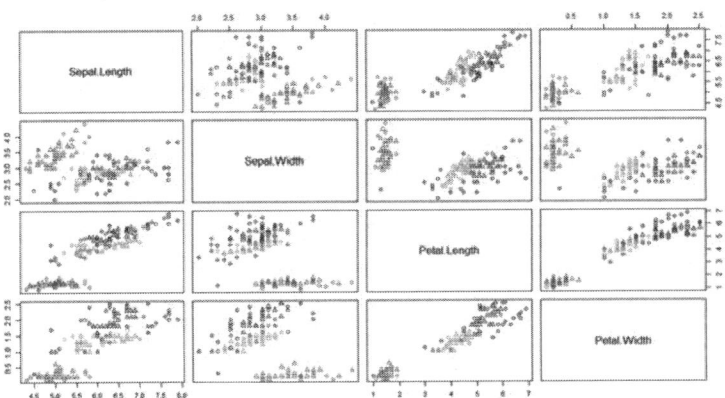

**Figure 6.28**   Plot of Outliers Detection Using Density Based Clustering

The clustering results can also be plotted in the scatter plot as in *Fig.* 6.29. This can be done using the function *plot()* in the *fpc* package. Here also the black circles denote the outliers.

> plot(db, iris1[c(1,2)])

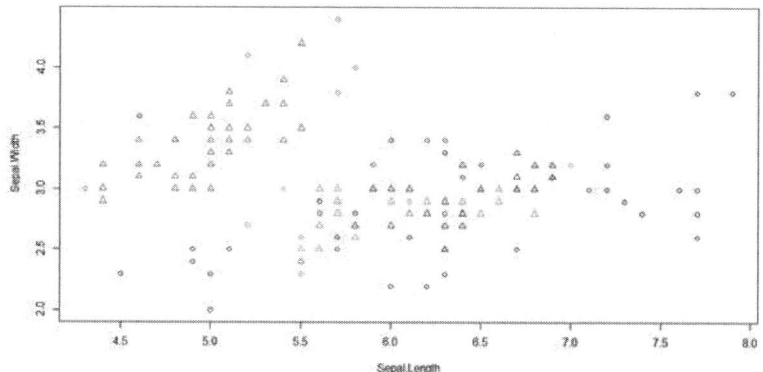

**Figure 6.29** Scatter Plot of Outliers Detection Using Density Based Clustering

We can also detect outliers with the *k-means* algorithm. With *k-means*, the data are partitioned into k groups by assigning them to the closest cluster centres. After that, we can calculate the distance (or dissimilarity) between each object and its cluster centre, and pick those with largest distances as outliers. An example of outlier detection with *k-means* from the *iris* data is given below. In the graph of *Fig. 6.30*, the cluster centres are labelled with "*" and outliers with "+".

```
> iris2 <- iris[,1:4]
> km <- kmeans(iris2, centers=3)
> centers <- km$centers[km$cluster, ]
> dist <- sqrt(rowSums((iris2 - centers) ^ 2))
> outliers <- order(dist, decreasing=T)[1:5]

> print(outliers)
[1] 99 58 94 61 119

> print(iris2[outliers,])
```

|    | Sepal.Length | Sepal.Width | Petal.Length | Petal.Width |
|----|--------------|-------------|--------------|-------------|
| 99 | 5.1          | 2.5         | 3.0          | 1.1         |
| 58 | 4.9          | 2.4         | 3.3          | 1.0         |

| 94 | 5.0 | 2.3 | 3.3 | 1.0 |
| 61 | 5.0 | 2.0 | 3.5 | 1.0 |
| 119 | 7.7 | 2.6 | 6.9 | 2.3 |

> plot(iris2[,c("Sepal.Length", "Sepal.Width")], pch="o", col=km$cluster, cex=0.3)
> points(km$centers[,c("Sepal.Length", "Sepal.Width")], col=1:3, pch=8, cex=1.5)
> points(iris2[outliers, c("Sepal.Length", "Sepal.Width")], pch="+", col=4, cex=1.5)

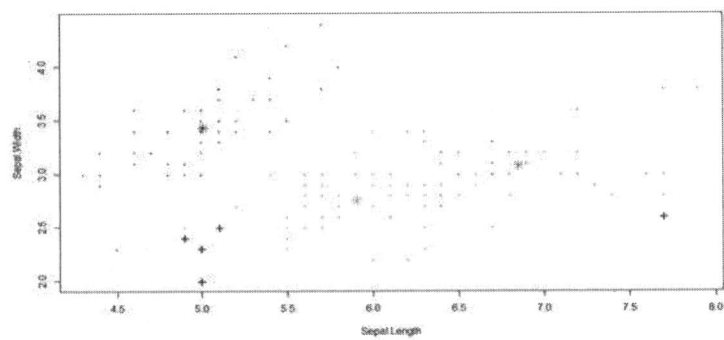

**Figure 6.30**   Scatter Plot of Outliers Detection Using K-Means Clustering

## 6.6. Dimensionality Reduction Using R

Dimensionality reduction is the process of reducing the number of irrelevant variables in a given dataset for future predictive modelling. It is a very important step of predictive modelling. If the idea is to improve accuracy of the model, it's the step where one needs to invest more time. Right variables produce more accurate models. Hence, it is required to choose a strategy for variable selection for efficient prediction from the given data. Some of these strategies are Principal Component Analysis, Singular Value Decomposition and Feature Selection.

### 6.6.1. Principal Component Analysis

Sparse data is a commonly faced problems in data analytics. At many times, we face a situation where we have a large set of features and fewer data points, or we have data with very high feature vectors. In such scenarios, fitting a model to the dataset results in lower predictive power of the model. This scenario is often termed as the curse of dimensionality. *Principal Component Analysis (PCA)* is a popular dimensionality reduction technique. PCA is used in many applications that deals with high dimension of data.

Consider the data, we want to work with, is in the form of a matrix A of $m \times n$ dimensions, where $A_{i,j}$ represents the value of the $i^{th}$ observation of the $j^{th}$ variable. Thus the $n$ members of the matrix can be identified with the $m$ rows, each variable corresponding to $n$-dimensional vectors. If $n$ is very large it is often desirable to reduce the number of variables to a smaller number of variables, say $k$ variables, where $k < n$, while losing as little information as possible. PCA is a *linear orthogonal transformation* that transforms the data to a new coordinate system. The projection of the data lies on the first coordinate called as the first principal component. The second transformation happens on the second coordinate called the second principal component, and so on. The variables that have less variance are discarded by the PCA technique.

PCA uses orthogonal projection of highly correlated variables to a set of values of linearly uncorrelated variables called *Principal Components*. The number of reduced principal components / features is less than or equal to the number of original variables / features. The transformation is done in such a way that the first principal component has the largest variance. By considering the highly correlated features, the data varies as much as possible. Every next component has the highest variance. This uses the features that are less correlated with the first principal component and that are orthogonal to the preceding component.

We can implement *PCA* in R by using the *crimtab* dataset available in R. This dataset contains the data of 3000 *male criminals* over 20 *years* old undergoing their sentences in the chief prisons of England and Wales. The row names correspond to midpoints of intervals of finger lengths whereas the column names correspond to heights of 3000 criminals.

> *str(crimtab)*

*'table' int [1:42, 1:22] 0 0 0 0 0 0 1 0 0 0 ...*

*- attr(\*, "dimnames")=List of 2*

*..$ : chr [1:42] "9.4" "9.5" "9.6" "9.7" ...*

*..$ : chr [1:22] "142.24" "144.78" "147.32" "149.86" ...*

*> dim(crimtab)*

*[1] 42 22*

*> sum(crimtab)*

*[1] 3000*

*> nrow(crimtab)*

*[1] 42*

*> colnames(crimtab)*

*[1] "142.24" "144.78" "147.32" "149.86" "152.4" "154.94" "157.48" "160.02"*
  *"162.56" "165.1"*

*[11] "167.64" "170.18" "172.72" "175.26" "177.8" "180.34" "182.88" "185.42"*
  *"187.96" "190.5"*

*[21] "193.04" "195.58"*

*> rownames(crimtab)*

*[1] "9.4" "9.5" "9.6" "9.7" "9.8" "9.9" "10" "10.1" "10.2" "10.3" "10.4" "10.5"*
                                                                 *"10.6"*

*[14] "10.7" "10.8" "10.9" "11" "11.1" "11.2" "11.3" "11.4" "11.5" "11.6" "11.7"*
                                                        *"11.8" "11.9"*

*[27] "12" "12.1" "12.2" "12.3" "12.4" "12.5" "12.6" "12.7" "12.8" "12.9" "13"*
                                                        *"13.1" "13.2"*

*[40] "13.3" "13.4" "13.5"*

Let us use the function *apply()* to the *crimtab* dataset row wise to calculate the variance to see how each variable is varying. The function *apply()* returns a vector or array or list of values obtained by applying a function to margins of an array or matrix.

*> apply(crimtab,2,var)*

| 142.24 | 144.78 | 147.32 | 149.86 | 152.4 |
|---|---|---|---|---|
| 0.02380952 | 0.02380952 | 0.17421603 | 0.88792102 | 2.56445993 |

| 154.94 | 157.48 | 160.02 | 162.56 | 165.1 |
|--------|--------|--------|--------|-------|
| 11.19860627 | 38.04471545 | 107.76596980 | 205.79616725 | 270.58536585 |
| 167.64 | 170.18 | 172.72 | 175.26 | 177.8 |
| 238.42973287 | 187.60569106 | 79.57491289 | 42.85540070 | 12.51161440 |
| 180.34 | 182.88 | 185.42 | 187.96 | 190.5 |
| 3.74680604 | 0.68583043 | 0.19105691 | 0.08826945 | 0.00000000 |
| 193.04 | 195.58 | | | |
| 0.00000000 | 0.02380952 | | | |

We can see that the column "165.1" contains maximum variance "270.58536585". Let us now apply PCA using the function prcomp() on the data set crimtab.

```
> pca =prcomp(crimtab)
> pca
```

*Standard deviations (1, .., p=22):*

[1] 30.07962021 14.61901911 5.45438277 4.65250574 3.21408168 2.77322835
                                                                2.30250353

[8] 1.92678188 1.40986049 1.24320894 1.02967875 0.72502776 0.50683548
                                                                0.47841947

[15] 0.29167315 0.26636232 0.22462458 0.12793888 0.12483426 0.06548509
                                                                0.00000000

[22] 0.00000000

*Rotation (n x k) = (22 x 22):*

|  | PC1 | PC2 | PC3 | PC4 | PC5 | PC6 |
|--|-----|-----|-----|-----|-----|-----|
| 142.24 | 0.0006121303 | -9.156788e-05 | 0.0019577283 | -0.002856530 | 0.0009242447 | - 0.002241331 |
| 144.78 | 0.0005879745 | -4.169480e-04 | -0.0012002251 | -0.001636190 | 0.0037931852 | -0.00111885 |
| ... | | | | | | |
| ... | | | | | | |

From the above code, the resultant components of PCA object are the standard deviations and Rotation. From the standard deviations we can observe that the 1st PCA explained most of the variation, followed by other PCAs'. The proportion of each variable along each principal component is given by the Rotation of the principal component. Let's plot all the principal components and see how the variance is accounted with each component in Fig. 6.31.

```
> par(mar = rep(2, 4))
> plot(PCA)
```

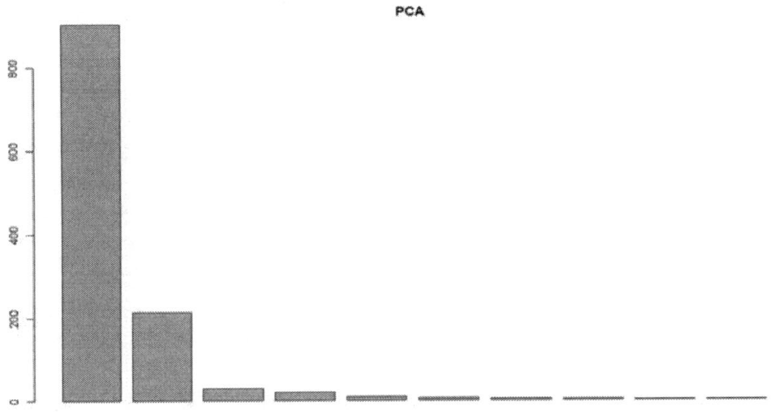

PCA

**Figure 6.31**   Plot of Variances of the Principal Components

Clearly the first principal component accounts for maximum information. The results of PCA can be represented as a *biplot* graph. *Biplot* is used to show the proportions of each variable along the first two principal components as in *Fig. 6.32*. The first two lines of the below code changes the direction of the *biplot*. If we do not include the first two lines the plot will be mirror image of the below graph.

```
> PCA$rotation=-PCA$rotation
> PCA$x=-PCA$x
> biplot (PCA, scale = 0.2)
```

The below *Fig. 6.32*, is known as a *biplot*. In this, we can see the two principal components (*PC1* and *PC2*) of the *crimtab* dataset ploted in the graph. The arrows represent the loading vectors, and this specifies how the feature space varies along the principal component vectors. From the below plot, we can see that the first principal component vector, *PC1*, more or less places equal weight on three features: *165.1*, *167.64*, and *170.18*. This means that these three features are more correlated with each other. In the second principal component, *PC2* places more weight on *160.02*, *162.56* than the other 3 features.

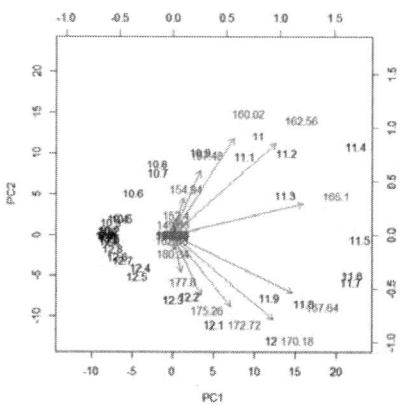

**Figure 6.32**    Biplot of the First Two Principal Components

## 6.6.2. Feature Selection

*"Feature Selection"* or *"Variable Selection "* is a process of reducing the number of variables that are irrelevant.

### Step I: Remove Redundant Variables

The following 3 simple analysis helps to remove redundant variables:

1.  Remove variables having high percentage of missing values
2.  Remove Zero and Near Zero-Variance Predictors
3.  Remove highly correlated variables (greater than 0.7)

The absolute values of pair-wise correlations are considered. Then, we remove the variable with the largest mean absolute correlation looking at the mean absolute correlation of each variable, if the two variables have high correlation. The below code in R does the above steps.

```
> library(caret)

> library(corrplot)

> library(plyr)

> dat <- read.csv("Sample.csv")

> set.seed(227)

# Remove variables having high percentage of missing values

> dat1 <- dat[, colMeans(is.na(dat)) <= .5]

> dim(dat1)

[1] 19622   93

> dim(dat)

[1] 19622   160

> nzv <- nearZeroVar(dat1)

# Remove Zero and Near Zero-Variance Predictors

> dat2 <- dat1[, -nzv]

> dim(dat2)

[1] 19622   59

> numericData <- dat2[sapply(dat2, is.numeric)]

> summary(descrCor[upper.tri(descrCor)])
```

| Min. | 1st Qu. | Median | Mean | 3rd Qu. | Max. |
|------|---------|--------|------|---------|------|
| -0.992008 | -0.101969 | 0.001729 | 0.001405 | 0.084718 | 0.980924 |

```
> corrplot(descrCor, order = "FPC", method = "color", type = "lower",
                                        tl.cex = 0.7, tl.col = rgb(0, 0, 0))
```

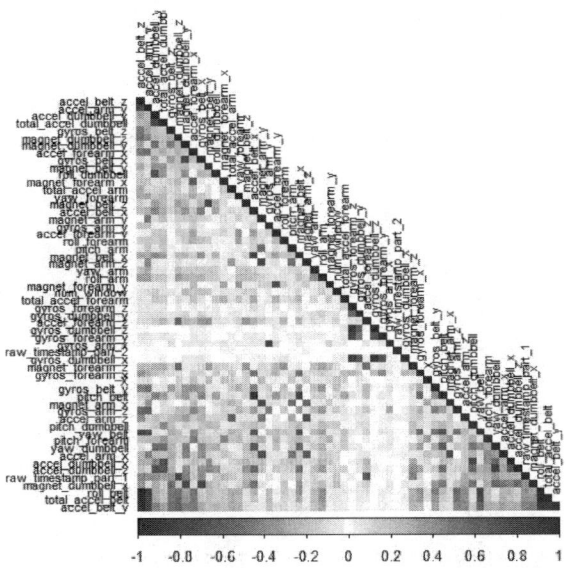

**Figure 6.33 – Correlation Plot of the Features**

> *highlyCorrelated <- findCorrelation(descrCor, cutoff=0.7)*

> *print(highlyCorrelated)*

[1] 14  5 13 26  8  7 40 12 15 41 39 42 43 25 38 27 29 16 52 50 49 37 22

> *highlyCorCol <- colnames(numericData)[highlyCorrelated]*

# *Remove highly correlated variables (greater than 0.7)*

> *dat3 <- dat2[, -which(colnames(dat2) %in% highlyCorCol)]*

> *dim(dat3)*

[1] 19622   36

### Step II: Feature Selection with Random Forest

Random Forest is one of the most popular algorithm in data science. The best part of this algorithm is there are no assumptions attached to it. The pre-processing work is very less as compared to other techniques. It overcomes the problem of over fitting that decision tree has. It provides a list of predictor (independent) variables which are important in predicting the target (dependent) variable. It contains two measures of variable importance. The first one - Gini Gain produced by the variable, averaged over all trees. The second one - Permutation Importance i.e. mean decrease in classification accuracy after permuting the variable, averaged over all trees. Sort the permutation importance score on descending order and select the *TOP k* variables. The below code is for feature selection with Random Forest.

```
> library(randomForest)
> rf <-randomForest(classe~.,data=dat3, importance=TRUE,ntree=100)
# Finding importance of variables
> imp = importance(rf, type=1)
> imp <- data.frame(predictors=rownames(imp),imp)
# Sorting the variables in descending order of their MeanDecreaseAccuracy
> imp.sort <- arrange(imp,desc(MeanDecreaseAccuracy))
> imp.sort$predictors <-
          factor(imp.sort$predictors,levels=imp.sort$predictors)
> imp.20<- imp.sort[1:20,]
# Printing top 20 variables with high MeanDecreaseAccuracy
> print(imp.20)
```

|    | predictors | MeanDecreaseAccuracy |
|----|------------|----------------------|
| 1  | X | 36.878224 |
| 2  | raw_timestamp_part_1 | 19.939217 |
| 3  | cvtd_timestamp | 19.936367 |
| 4  | pitch_belt | 14.474235 |
| 5  | roll_dumbbell | 12.502391 |
| 6  | gyros_belt_z | 12.429689 |
| 7  | num_window | 11.491461 |
| 8  | total_accel_dumbbell | 11.193014 |
| 9  | gyros_arm_y | 10.509349 |
| 10 | magnet_forearm_z | 10.353922 |

| 11 | *gyros_dumbbell_x* | 10.245442 |
| 12 | *magnet_belt_z* | 10.078787 |
| 13 | *pitch_forearm* | 10.069103 |
| 14 | *roll_arm* | 10.049374 |
| 15 | *yaw_arm* | 9.959173 |
| 16 | *gyros_dumbbell_y* | 9.770771 |
| 17 | *gyros_belt_x* | 9.602383 |
| 18 | *magnet_forearm_y* | 9.407758 |
| 19 | *user_name* | 9.304626 |
| 20 | *gyros_forearm_x* | 8.954952 |

*> varImpPlot(rf, type=1)*

*# Retaining only top 20 variables with high MeanDecreaseAccuracy in dat4*

*> dat4 = cbind(classe = dat3$classe, dat3[,c(imp.20$predictors)])*

To fix the number of features to be selected based on the ranking, a threshold (say '*n*') is fixed and based on this the top n (20) features are selected for further analysis.

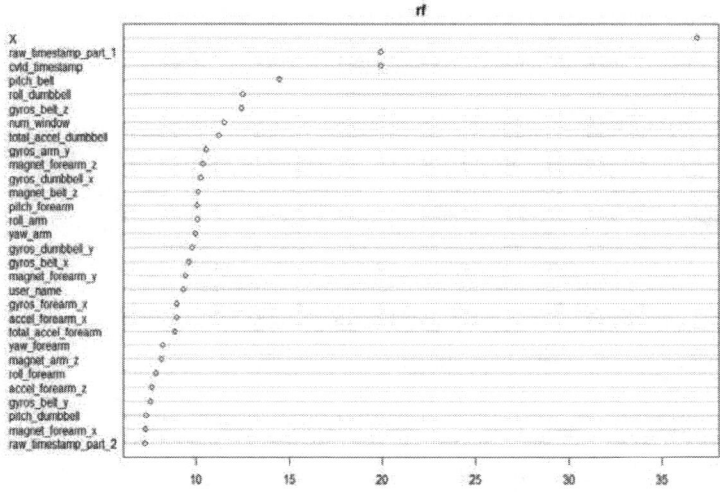

**Figure 6.34** Correlation Plot of the Features

❖ **HIGHLIGHTS**

- The R packages that are related to data mining are *stats, cluster, fpc, sna, e1071, cba, biclust, clues, kohonen, rpart, party, randomForest, ada, caret, arules, eclat, arulesViz, DMwR, dprep, Rlof, plyr, corrplot, RWeka, gausspred, optimsimplex, CCMtools, FactoMineR* and *nnet*.

- The R packages and function for clustering are *stats - kmeans(), cluster - pam(), fpc - pamk(), cluster - agnes(), stats - hclust(), cluster - daisy(), fpc - dbscan(), sna - kcores(), e1071 - cmeans(), cba - rockCluster(), biclust - biclust(), clues - clues(), kohonen - som(), cba - proximus()* and *cluster - clara()*.

- The functions *kmeans(), table(), plot()* and *point()* are used below for getting and plotting the results of K-Means clustering.

- R has the *pam()* and *pamk()* functions of the *cluster* package to do the k-medoids clustering.

- The *silhouette width* decides the number of clusters in k-medoids clustering.

- The hierarchical clustering in R can be done using the function *hclust()* in the *fpc* package.

- The function *dbscan()* in the package *fpc* is used for the density based clustering.

- The R packages and function for clustering are *e1071 - svm(), RWeka - IBk(), party - rpart(), rpart - ctree(), rpart - cforest(), randomForest - randomForest(), e1071 - naiveBayes(), ada - ada(), caret - train()*.

- The function *ctree()* in the package *party* and the function *rpart()* in the package *rpart* can be used to build the decision tree for the given data.

- We can do the prediction using the function *predict()* for the given new data.

- The function *randomForest()* in the package *randomForest* can be used to classify the given data.

- The importance of variables in a dataset can be obtained with the functions *importance()* and *varImpPlot()*.

- The built random forest is tested on test data, and the result is checked with the functions *table()* and *margin()*.

- An implementation of Association Rule Mining using APRIORI algorithm is the function *apriori()* in package *arules*.

- An implementation of Association Rule Mining using ECLAT algorithm is the function *eclat()* in package *eclat*.

- The function *inspect()* can be used to view the rules generated by the function *apriori()*.

- The package *arulesViz* is used to visualize the association rules as graphs and plots.

- Univariate outlier detection is done with function *boxplot.stats()*.

- Function *lofactor()* available in the packages *DMwR* and *dprep* calculates local outlier factors using the LOF algorithm.

- The function *prcomp()* performs a principal component analysis, and *biplot()* plots the data with its first two principal components.

- Package *Rlof* provides the function *lof()*, a parallel implementation of the LOF algorithm.

# CHAPTER 7

# CASE STUDIES

## ❖ OBJECTIVES

On completion of this Chapter you will be able to:

- understand how text mining is done using R
- understand how modelling can be done using R for credit risk analysis
- understand how EDA can be conducted on the two data sets, the crime data and the base ball data
- understand how social network analysis can be done using R

## 7.1. Text Mining

Text mining, is the process of deriving knowledge / insights from textual data. High-quality information is typically derived through the devising of patterns and trends through pattern learning. Text mining involves the process of preprocessing / formatting the input text, finding patterns within the preprocessed data, and finally evaluation of the results. The tasks of text mining are categorization of text, clustering of text, concept / entity extraction, sentiment analysis, production of granular taxonomies, entity relation modelling and document summarization.

This case study on text mining starts with using the twitter feeds from the dataset "*GameReview.csv*" for further analysis. The extracted text is then transformed to build a document-term matrix. Then, frequent words and associations are found from the matrix. Important words in a document can be presented as a word cloud. Packages used for text mining are "*tm*" and "*wordcloud*".

> *library(tm)*

```
> dat<-read.csv("GameReview.csv",stringsAsFactors = FALSE)

> dat[1:5,4]
```

[1] " SO ADDICTING  DEFF DOWNLAOD ITS EPIC YOU CAT LOVERS WILL
     FALL IN LOVE <3"

[2] " Great game I love this game. Unlike other games they constantly give you money to
play. They are always given you a bone. Keep up the good work."

[3] " Sooo much FUN I would definitely recommend this game, it's fun for dress up and
business. It's extremely entertaining, I'm hooked already. :)"

[4] " AWESOME Epic game so addictive 5stars <f0><U+009F><U+0098>
     <U+0084>"

[5] " Good game Great game"

The data is first converted to a corpus, which is a collection of text documents. After that, the corpus can be processed with functions provided in package "tm".

```
> corp <- Corpus(VectorSource(dat$text))
```

The corpus needs a couple of transformations, including changing letters to lower case, and removing punctuations, numbers and stop words. The general English stop-word list is tailored by adding and removing few words.

```
> corp <- tm_map(corp, content_transformer(tolower))

> corp <- tm_map(corp, removePunctuation)

> corp <- tm_map(corp, stripWhitespace)

> corp <- tm_map(corp, removeWords, stopwords("english"))
```

In the above code, the function *tm_map()* is an interface to apply transformations to the corpus. Word stemming can be done and after that, the stems can be completed to their original forms, so that the words would look normal. This can be achieved with function *stemCompletion()*. Before applying stemming a copy of the corpus is taken to be used as dictionary for stem completion.

```
> corpcopy <- corp

> corp <- tm_map(corp, stemDocument)

> for (i in 1:5) {
```

```
> cat(paste("[[", i, "]] ", sep=""))
> writeLines(strwrap(corp[[i]], width=60))
> }
```

[[1]] addict deff downlaod epic cat lover will fall love 3

[[2]] great game love game unlik game constant give money play

alway given bone keep good work

[[3]] sooo much fun definit recommend game fun dress busi extrem

entertain im hook alreadi

[[4]] awesom epic game addict 5star f0u009fu0098u0084

[[5]] good game great game

After that, we use *stemCompletion()* to complete the stems with the unstemmed corpus *corpcopy* as a dictionary. With the default setting, it takes the most frequent match in dictionary as completion.

```
> stemCompletion2 <- function(x, dictionary) {
+    x <- unlist(strsplit(as.character(x), " "))
+    x <- x[x != ""]
+    x <- stemCompletion(x, dictionary=dictionary)
+    x <- paste(x, sep="", collapse=" ")
+    stripWhitespace(x)
+ }
> corp <- lapply(corp, stemCompletion2, dictionary=corpcopy)
> corp <- Corpus(VectorSource(corp))
> for (i in 1:5) {
+    cat(paste("[[", i, "]] ", sep=""))
+    writeLines(strwrap(corp[[i]], width=60))
+ }
```

[[1]] addicted deff downlaod epic cat lovers will fall love 3

[[2]] great game love game unlike game constant give money play *always given bone keep good work*

*[[3]] sooo much fun definitely recommend game fun dress business*

*extremely entertained im hooked*

*[[4]] awesome epic game addicted 5stars f0u009fu0098u0084*

*[[5]] good game great game*

From the listings from the corpus before stemming, after stemming and after stemming completion we can see the words being changed. For example in line 4, the word was *"addictive"* before stemming, the word became *"addict"* after stemming and then after stemming completion it became *"addicted"*.

The relationship between terms and documents are represented in a *term-document matrix*, where each row stands for a term and each column stands for a document. Each entry in this matrix is the number of occurrences of the term in the document. Alternatively, one can also build a *document-term matrix* by swapping row and column. We then build a term-document matrix from the above processed corpus with function *TermDocumentMatrix()*. With its default setting, terms with less than three characters are discarded.

> *tdm <- TermDocumentMatrix(corp)*

> *tdm*

   *<<TermDocumentMatrix (terms: 2117, documents: 1000)>>*

| | |
|---|---|
| *Non-/sparse entries* | *: 11144/2105856* |
| *Sparsity* | *: 99%* |
| *Maximal term length* | *: 391* |
| *Weighting* | *: term frequency (tf)* |

As we can see from the above result, the term-document matrix is composed of 2117 terms and 1000 documents. It is very sparse, with 99% of the entries being zero. We then have a look at the first six terms starting with "g" and tweets numbered 201 to 210.

> *idx = grep(glob2rx("g*"), dimnames(tdm)$Terms)*

> *inspect(tdm[idx,201:210])*

   *<<TermDocumentMatrix (terms: 81, documents: 10)>>*

| | | | | | | | | | | |
|---|---|---|---|---|---|---|---|---|---|---|
| Non-/sparse entries | | : 12/798 | | | | | | | | |
| Sparsity | | : 99% | | | | | | | | |
| Maximal term length | | : 72 | | | | | | | | |
| Weighting | | : *term frequency (tf)* | | | | | | | | |
| Sample | | : | | | | | | | | |

| | Docs | | | | | | | | | |
|---|---|---|---|---|---|---|---|---|---|---|
| Terms | 201 | 202 | 203 | 204 | 205 | 206 | 207 | 208 | 209 | 210 |
| game | 1 | 2 | 1 | 1 | 0 | 0 | 0 | 1 | 2 | 0 |
| gem | 0 | 2 | 0 | 0 | 0 | 0 | 0 | 0 | 0 | 0 |
| generally | 0 | 0 | 0 | 0 | 0 | 0 | 0 | 0 | 0 | 0 |
| get | 0 | 1 | 0 | 0 | 0 | 0 | 0 | 2 | 0 | 0 |
| give | 0 | 0 | 0 | 0 | 0 | 0 | 0 | 0 | 0 | 0 |
| given | 0 | 0 | 0 | 0 | 0 | 0 | 0 | 0 | 0 | 0 |
| glad | 0 | 0 | 0 | 0 | 0 | 0 | 0 | 0 | 0 | 0 |
| good | 0 | 0 | 0 | 1 | 0 | 0 | 0 | 0 | 0 | 0 |
| graphic | 0 | 0 | 0 | 0 | 0 | 0 | 0 | 0 | 0 | 0 |
| great | 0 | 2 | 0 | 0 | 0 | 0 | 0 | 0 | 1 | 0 |

Many data mining tasks can be done, based on the above matrix. For example, clustering, classification and association rule mining. When there are too many terms, the size of a term-document matrix can be reduced by selecting terms that appear in a minimum number of documents.

We will now have a look at the popular words and the association between words from the 1000 tweets.

```
> findFreqTerms(tdm, lowfreq=80)
[1] "addicted" "love"    "game"    "good"    "great"    "play"    "fun"
[8] "awesome"  "get"     "time"    "like"    "just"     "update"  "app"
[15] "can"     "cant"
```

In the code above, *findFreqTerms()* finds frequent terms with frequency no less than eighty. Note that they are ordered alphabetically, instead of by frequency or popularity. To show the top frequent words visually, we next make a *bar plot* for them. From the *term document matrix*, we can derive the frequency of terms with *rowSums()*. Then we select terms that appears in eighty or more documents and shown them with a *bar plot* using the package *ggplot2*. In the code below, *geom="bar"* specifies a *bar plot* and *coord_flip()* swaps x-axis and y-axis. The *bar plot* below clearly shows that the three most frequent words are *"game", "play"* and *"great"*.

```
> termFrequency <- rowSums(as.matrix(tdm))

> termFrequency <- subset(termFrequency, termFrequency>=80)

> library(ggplot2)

> df <- data.frame(term=names(termFrequency), freq=termFrequency)

> ggplot(df, aes(x=term, y=freq)) + geom_bar(stat="identity") +

+    xlab("Terms") + ylab("Count") + coord_flip()
```

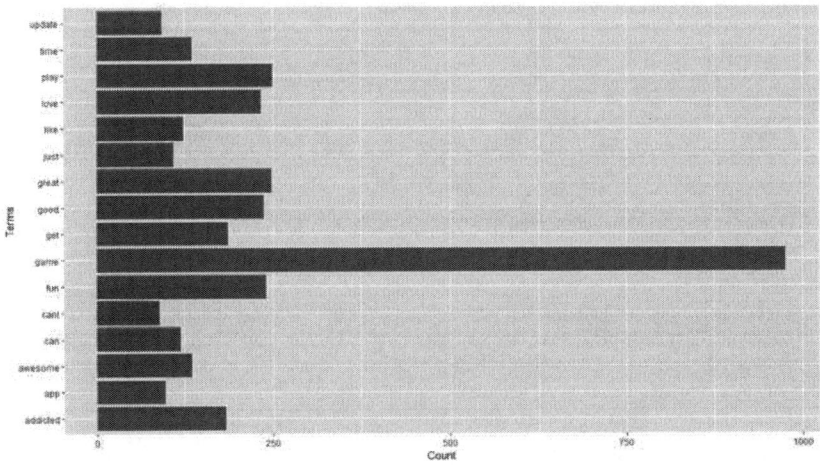

**Figure 7.1**   Ggplot of Terms and Frequencies

Alternatively, the above plot can also be drawn with *barplot()* as below, where the argument *las* sets the direction of x-axis labels to be vertical.

```
> barplot(termFrequency, las=2)
```

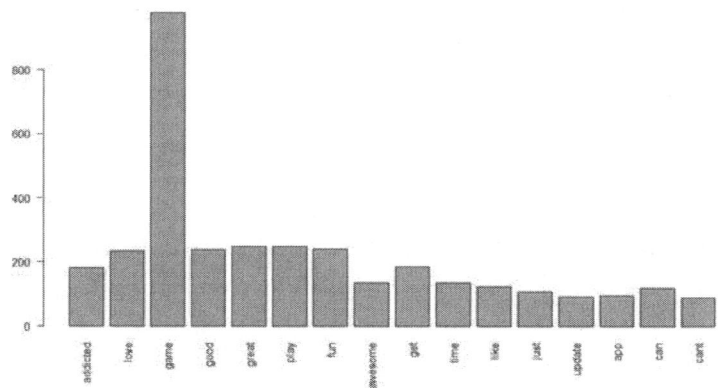

**Figure 7.2** Bar Plot of Terms and Frequencies

It is also possible to find the highly associated words with another word with the function *findAssocs()*. Below is the code to find the terms associated with the words *"game"* and *"play"* with correlation no less than 0.20 and 0.25 respectively. The words are ordered by their correlation with the terms *"game"* (or *"play"*).

```
> findAssocs(tdm, "game", 0.20)
```

*$game*

| play | player | dont | say | new | thatd | ever |
|------|--------|------|-----|-----|-------|------|
| 0.29 | 0.22 | 0.21 | 0.21 | 0.21 | 0.21 | 0.20 |

```
> findAssocs(tdm, "play", 0.25)
```

*$play*

| game | potential | course | dont | year | will | meter |
|------|-----------|--------|------|------|------|-------|
| 0.29 | 0.27 | 0.27 | 0.26 | 0.26 | 0.25 | 0.25 |

After building a term-document matrix, the importance of words can be shown with a word cloud, which can be easily produced with package wordcloud. The code below, first converts the term-document matrix to a normal matrix, and then calculates the word frequencies. After that, gray levels are set based on word frequency and we use the function wordcloud() to make a plot for it. With wordcloud(), the first two parameters give a list of words and their frequencies.

Words with frequency below twenty are not plotted, as specified by min.freq=20. By setting random.order=F, frequent words are plotted first, which makes them appear in the centre of cloud. We also set the colours to gray levels based on frequency.

> *library(wordcloud)*

> *m <- as.matrix(tdm)*

> *wordFreq <- sort(rowSums(m), decreasing=TRUE)*

> *pal <- brewer.pal(9, "BuGn")*

> *pal <- pal[-(1:4)]*

> *set.seed(375)*

> *grayLevels <- gray( (wordFreq+10) / (max(wordFreq)+10) )*

> *wordcloud(words=names(wordFreq), freq=wordFreq, min.freq=20,*

> *random.order=F, colors=pal)*

**Figure 7.3**   Word Cloud based on Frequencies

The above word cloud clearly shows again that *"game"*, *"play"* and *"great"* are the top three words, which validates that the Game review tweets. Some other important words are *"love"*, *"good"* and *"fun"*, which shows that the review on the game is very good.

With hierarchical clustering, we find clusters of words. The plot of clustering should not be crowded with words and hence we remove the sparse terms. Then the distances between terms are calculated with *dist()* after scaling. After that, the terms are clustered with *hclust()* and the *dendrogram* is cut into 4 clusters. The agglomeration method is set to ward, which denotes the increase in variance when two clusters are merged.

> *tdm2 <- removeSparseTerms(tdm, sparse=0.90)*

> *m2 <- as.matrix(tdm2)*

> *distMatrix <- dist(scale(m2))*

> *fit <- hclust(distMatrix, method="ward.D")*

> *plot(fit)*

> *rect.hclust(fit, k=4)*

> *(groups <- cutree(fit, k=4))*

| addicted | love | game | good | great | play | fun | awesome | get |
|:---:|:---:|:---:|:---:|:---:|:---:|:---:|:---:|:---:|
| 1 | 1 | 2 | 3 | 4 | 1 | 4 | 1 | 1 |

*time*

1

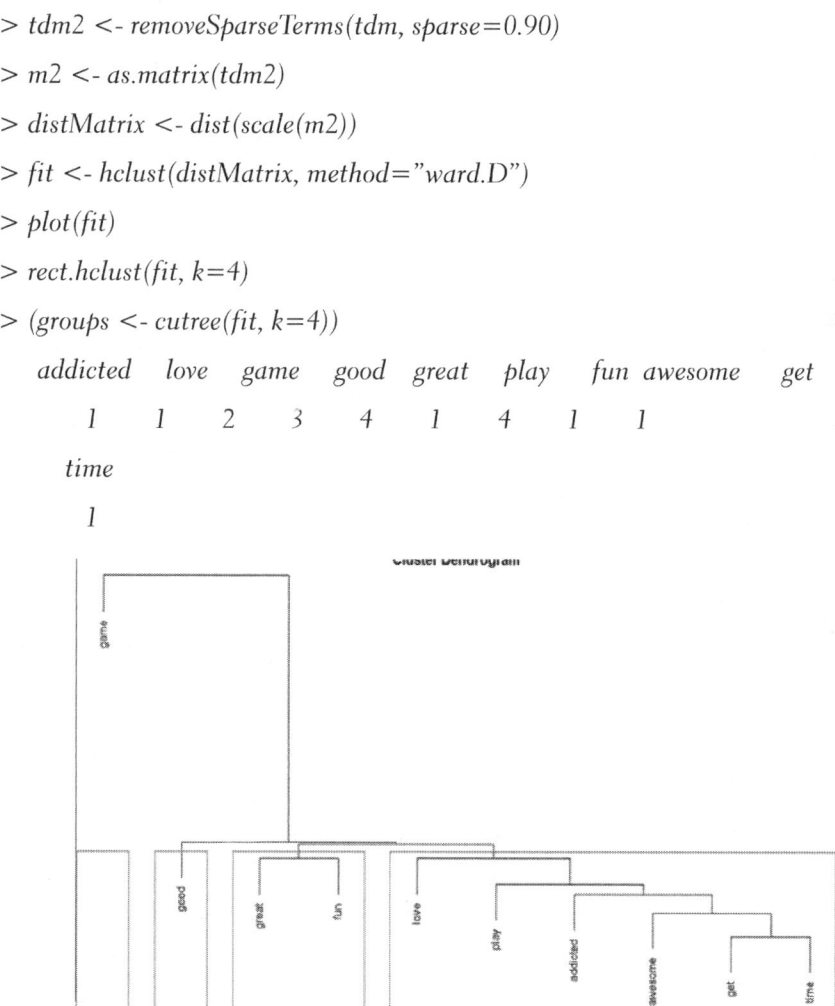

**Figure 7.4** Cluster of Words with High Frequencies Using Hierarchical Clustering

In the above *dendrogram*, we can see the topics in the tweets grouped into 4 different clusters. The most frequent word *"game"* is in the second cluster, the next frequent word "*good*" is in the third cluster, the words "*great*" and "*fun*" falls under the fourth cluster and the remaining words with low frequency falls under the first cluster.

Next we cluster the tweets using *k-means clustering* algorithm. The *k-means clustering* takes the values in the matrix as numeric. We transpose the *term-document matrix* to a document-term one. The tweets are then clustered with *kmeans()* with the number of clusters set to *eight*. After that, we check the popular words in every cluster and also the cluster centres. A fixed random seed is set with *set.seed()* before running *kmeans()*.

```
> m3 <- t(m2)
> set.seed(123)
> k <- 8
> kmeansResult <- kmeans(m3, k)
> round(kmeansResult$centers, digits=3)
```

|   | addicted | love | game | good | great | play | fun | awesome | get | time |
|---|----------|------|------|------|-------|------|-----|---------|-----|------|
| 1 | 0.168 | 0.107 | 1.267 | 0.137 | 1.313 | 0.069 | 0.176 | 0.130 | 0.145 | 0.061 |
| 2 | 0.227 | 0.394 | 3.652 | 0.409 | 0.348 | 0.803 | 0.273 | 0.258 | 0.409 | 0.258 |
| 3 | 0.038 | 0.017 | 1.174 | 0.547 | 0.000 | 0.127 | 0.068 | 0.127 | 0.081 | 0.106 |
| 4 | 0.212 | 0.061 | 1.091 | 0.333 | 0.030 | 0.606 | 0.212 | 0.030 | 2.242 | 0.788 |
| 5 | 0.183 | 0.113 | 1.507 | 0.113 | 0.183 | 0.887 | 1.085 | 0.197 | 0.197 | 0.254 |
| 6 | 1.205 | 0.154 | 0.397 | 0.077 | 0.090 | 0.154 | 0.423 | 0.179 | 0.026 | 0.064 |
| 7 | 0.000 | 0.118 | 0.000 | 0.125 | 0.104 | 0.146 | 0.212 | 0.090 | 0.066 | 0.097 |
| 8 | 0.237 | 1.381 | 1.216 | 0.031 | 0.031 | 0.216 | 0.062 | 0.175 | 0.134 | 0.093 |

To make it easy to find what the clusters are about, we then check the top three words in every cluster.

```
> for (i in 1:k) {
+    cat(paste("cluster ", i, ": ", sep=""))
+    s <- sort(kmeansResult$centers[i,], decreasing=T)
+    cat(names(s)[1:3], "\n")
+ }
```

*cluster 1: great game fun*

*cluster 2: game play good*

*cluster 3: game good play*

*cluster 4: get game time*

*cluster 5: game fun play*

*cluster 6: addicted fun game*

*cluster 7: fun play good*

*cluster 8: love game addicted*

From the above top words and centres of clusters, we can see that the clusters are of different topics. We can see in every cluster except the *cluster 7*, the word *"game"* is a part of it and each of these clusters talk about the game in different angle.

## 7.2. Credit Risk Analysis and Prediction

Credit risk evaluation has become more important nowadays for Banks to issue loans for their customers based on their credibility. For this the internal rating based approach is the most sought by the banks that need approval by the bank manager. The most accurate and highly used credit scoring measure is the *Probability of Default* called the *PD*. Defaulter is the one who is unlikely to repay the loan amount or will have overdue of loan payment by more than 90 days. Hence determining the PD is the crucial step for credit scoring of the customers seeking bank loan.

Hence a data mining framework is developed for PD estimation from a given set of data using the data mining techniques available in R Package. The data used to implement and test this model is taken from the *UCI Repository*. The *German credit scoring dataset with 1000 records and 21 attributes* is used for this purpose. The numeric format of the data is loaded into the R Software and a set of data preparation steps are executed before the same is used to build the classification model. The dataset that we have selected does not have any missing data. But, in real time there is possibility that the dataset has many missing or imputed data which needs to be replaced with valid data generated by making use of the available complete data. The k nearest neighbours algorithm is used for this purpose to perform multiple imputation. This is implemented using the *knnImputation()* function of package *DMwR*. The numeric features are normalized before this step.

The dataset has many attributes that define the credibility of the customers seeking for several types of loan. The values for these attributes can have outliers that do not fit into the regular range of data. Hence, it is required to remove the outliers before the dataset is used for further modelling. The outlier detection for quantitative features is done using the function *levels()*. For numeric features the *box plot* technique is used for outlier detection and this is implemented using the *daisy()* function of the *cluster* package. But, before this the numeric data has to be normalized into a domain of *[0, 1]*. The agglomerative hierarchical clustering algorithm is used for outlier ranking. This is done using the *outliers.ranking()* function of the *DMwR* package. After ranking the outlier data, the ones that is out of range is disregarded and the remaining outliers are filled with null values.

The inconsistencies in the data like unbalanced dataset have to be balanced before building the classification model. Many real time datasets have this problem and hence need to be rectified for better results. But, before this step, it is required to split the sample dataset into training and test datasets which will be in the ratio 4:1 (i.e. Training dataset 80% of data and 20% of data will be test dataset). Now the balancing step will be executed on the training dataset using the *SMOTE()* function of the *DMwR* package.

Next using the training dataset the correlation between the various attributes need to be checked to see if there are any redundant information represented using two attributes. This is implemented using the *plotcorr()* function in the *ellipse* package. The unique features will then be ranked and based on the threshold limit the number of highly ranked features will be chosen for model building. For ranking the features the *randomForest()* function of the *randomForest* package is used. The threshold for selecting the number of important features is chosen by using the *rfcv()* function of the *randomForest* package.

Now the resultant dataset with the reduced number of features is ready for use by the classification algorithms. Classification is one of the data analysis methods that predict the class labels. Classification can be done in several ways and one of the most appropriate for the chosen problem is using decision trees. Classification is done in two steps – (i) the class labels of the training dataset is used to build the decision tree model and (ii) This model will be applied on the test dataset to predict the class labels of the test dataset. For the first step the function *rpart()* of the *rpart* package will be used. The *predict()* function is used to execute the second step. The resultant prediction is then evaluated against the original class labels of the test dataset to find the accuracy of the model.

**Dataset Attribute Types**

| A1 | A2 | A3 | A4 | A5 | A6 | A7 | A8 | A9 | A10 | A11 | A12 | A13 | A14 | A15 | A16 | A17 | A18 | A19 | A20 | Def |
|----|----|----|----|----|----|----|----|----|-----|-----|-----|-----|-----|-----|-----|-----|-----|-----|-----|-----|
| Q | N | Q | Q | N | Q | Q | Q | Q | Q | Q | Q | N | Q | Q | N | Q | N | B | B | B |

*Q: Quantitative*   *N: Numeric*   *B: Binary*

A1: Status of Existing Account
   (1: < 0 DM, 2: < 200 DM, 3: > = 200 DM, 4: No existing Account)

A2: Loan Duration in Month

A3: Credit History
   (0: No credits taken so far, 1: All credit in this Bank paid back dully till now, 2: Existing credits paid back dully till now, 3: Delay in paying off in the past, 4: Credits existing in other banks)

A4: Loan Purpose
   (0: new car purchase, 1: used car purchase, 2: furniture or equipment purchase, 3: radio or television purchase, 4: domestic appliances purchase, 5: repairs, 6: education, 7: vacation, 8: retraining, 9: Business, 10: others)

A5: Credit Amount (in DM)

A6: Bonds / Savings
   (1: < 100 DM, 2: > = 100 and < 500 DM, 3: > = 500 DM and 1000 DM, 4: > = 1000 DM, 5: no savings / bonds)

A7: Present Employment Since
   (1: unemployed, 2: < 1 year, 3: > = 1 and < 4 years, 4: > = 4 and < 7 years, 5: > = 7 years)

A8: Instalment rate in percentage of disposable income

A9: Personal Status and Sex
   (1: Divorced Male, 2: Divorced/Married Female, 3: Male Single, 4: Married Male, 5: Female Single)

A10: Other Debtors / Guarantors
   (1: None, 2: Co-applicant, 3: Guarantor)

A11: Present Residence Since (in Years)

A12: Property

    (1: Real Estate, 2: Life Insurance, 3: Car or others, 4: No property)

A13: Age in years

A14: Other instalment plans

    (1: Bank, 2: Stores, 3: None)

A15: Housing

    (1: Rented, 2: Owned, 3: For Free)

A16: Number of existing credits at this bank

A17: Job Status

    (1: Unemployed non-resident, 2: Unemployed resident, 3: Skilled Employee, 4: Self-Employed )

A18: Number of People being liable to provide maintenance for

A19: Telephone

    (0: Not Available, 1: Available)

A20: Foreign Worker

    (0: No, 1: Yes)

Def: Class Label

    (0: Non Default, 1: Default)

## Dataset Selection

The *German Credit Scoring dataset* in the numeric format which is used for the implementation of this model has the below attributes and the descriptions of the same are given in the below table.

After selecting and understanding the dataset it is loaded into the R software using the below code. The dataset is loaded into R with the name *creditdata*.

> *creditdata <- read.csv("UCI German Credit Data Numeric.csv",*

> *header = TRUE, sep = ",")*

> *nrow(creditdata)*

*[1] 1000*

## Data Pre-Processing

1) *Outlier Detection:* To identify the outliers of the numeric attributes, the values of the numeric attributes are normalized into the domain range of *[0, 1]* and they are plotted as *box plot* to view the outlier values as in *Fig. 7.5*. The code and the result for this step are given as below.

> *normalization <- function(data,x)*

> *{for(j in x)*

> *{data[!(is.na(data[,j])),j]=*

> *(data[!(is.na(data[,j])),j]-min(data[!(is.na(data[,j])),j]))/*

> *(max(data[!(is.na(data[,j])),j])-min(data[!(is.na(data[,j])),j]))}*

> *return(data)}*

> *c <- c(2,5,8,11,13,16,18)*

> *normdata <- normalization(creditdata,c)*

> *boxplot(normdata[,c])*

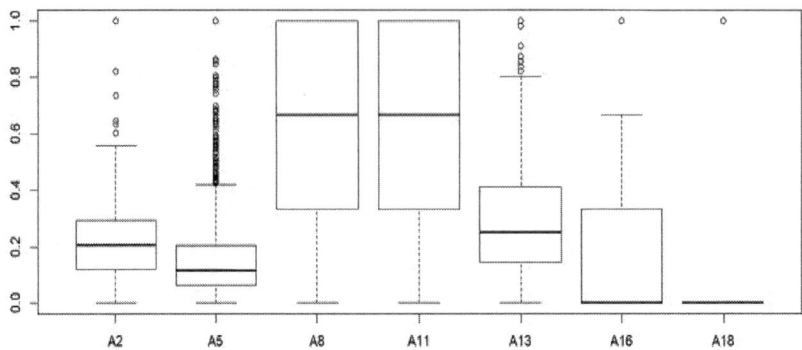

**Figure 7.5**   Box Plot of Outliers in Numeric Attributes

To identify the outliers of the quantitative attributes, the below commands are used. From the results of the same, one can identify the values that do not fall under the allowed values. For example the attribute "A1" can only take the values "1", "2", "3", "4". If there is any observation that has data other than these allowed values, it is removed. Similarly, the allowed values for each quantitative attribute can be checked and outliers removed.

> *levels(as.factor(creditdata[,"A1"]))*

*[1] "1" "2" "3" "4"*

2)   *Outliers Ranking:* The agglomerative hierarchical clustering algorithm chosen for ranking the outliers is less complex and easy to understand. Each observation is assumed to be a cluster and in each step, the observations are grouped based on the distance between them. Each observation that is observed later has lower rank. It is seen that the observations with lower rank are outliers because there are dissimilarities between them and the other observations. For outlier ranking the following code is used.

> *require(cluster)*

> *distance=daisy(creditdata[,-19],stand=TRUE,metric=c("gower"),*

   *type = list(interval=c(2,5,8,11,13,16,18),*

*nominal=c(1,3,4,6,7,9,10,12,14,15,17),binary=c(19,20)))*

> *require(DMwR)*

> *outlierdata=outliers.ranking(distance,test.data=NULL,method="sizeDiff",*

> *clus = list(dist="euclidean", alg = "hclust", meth="average"),*

> *power = 1, verb = F)*

3) *Outliers Removal: The observations which are out of range (based on the rankings) are removed using the below code. After outlier removal the dataset creditdata is renamed as creditdata_noout.*

> *boxplot(outlierdata$prob.outliers[outlierdata$rank.outliers])*

> *n=quantile(outlierdata$rank.outliers)*

> *n1=n[1]*

> *n4=n[4]*

> *filler=(outlierdata$rank.outlier > n4*1.3)*

> *creditdata_noout=creditdata[!filler,]*

> *nrow(creditdata_noout)*

*[1] 975*

4) *Imputations Removal: The method used for null values removal is multiple imputation method in which the k nearest neighbours' algorithm is used for both numeric and quantitative attributes. The numeric features are normalized before calculating the distance between objects. The following code is used for imputations removal. After imputations removal the dataset creditdata_noout is renamed as creditdata_noout_noimp.*

> *require(DMwR)*

> *creditdata_noout_noimp=knnImputation(creditdata_noout, k = 5, scale = T,*

> *meth = "weighAvg", distData = NULL)*

> *nrow(creditdata_noout_noimp)*

*[1] 975*

There were no null values for the attributes in the dataset we have chosen and hence the number of records remains unchanged after the above step.

5) *Splitting Training and Test Datasets: Before proceeding to the further steps, the dataset has to be split into training and test datasets so that the model can be built using the training dataset. The code for splitting the database is listed*

*below.*

> *library(DMwR)*

> *split<-sample(nrow(creditdata_noout_noimp),*

> *round(nrow(creditdata_noout_noimp)\*0.8))*

> *trainingdata=creditdata_noout_noimp[split,]*

> *testdata=creditdata_noout_noimp[-split,]*

6)   *Balancing Training Dataset: The SMOTE() function handles unbalanced classification problems and it generates the new smoted dataset that addresses the unbalanced class problem. It artificially generates observations of minority classes using the nearest neighbours of this class of elements to balance the training dataset. The following code is used for balancing the training dataset.*

> *creditdata_noout_noimp_train=trainingdata*

> *creditdata_noout_noimp_train$default <-*

*factor(ifelse(creditdata_noout_noimp_train$Def == 1, "def", "nondef"))*

> *creditdata_noout_noimp_train_smot <-*

*SMOTE(default ~ ., creditdata_noout_noimp_train, k=5,perc.over = 500)*

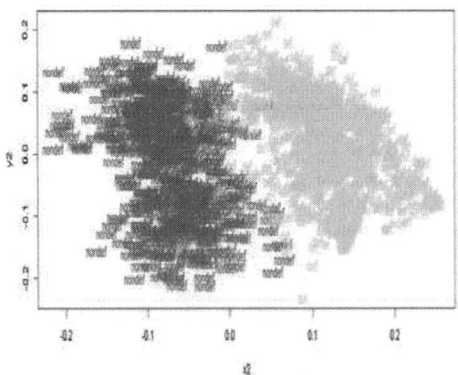

**Figure 7.6**   Data Distribution before Balancing

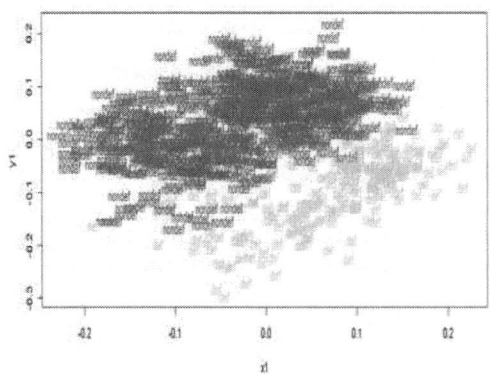

**Figure 7.7** Data Distribution after Balancing

The data distribution before and after balancing the data are shown in the figures *Fig. 7.6* and *Fig. 7.7*. This method is based on proximities between objects and produces a spatial representation of these objects. Proximities represent the similarity or dissimilarity between data objects. The code used to plot these objects is shown below.

```
> library(cluster)
> dist1=daisy(creditdata_noout_noimp_train[,-21],stand=TRUE,metric=c("gower"),
                         type = list(interval=c(2,5,8,11,13,16,18),
              nominal=c(1,3,4,6,7,9,10,12,14,15,17),binary=c(19,20)))
> dist2=daisy(creditdata_noout_noimp_train_smot[,-21],
stand=TRUE,metric=c("gower"),
                         type = list(interval=c(2,5,8,11,13,16,18),
              nominal=c(1,3,4,6,7,9,10,12,14,15,17),binary=c(19,20)))
> loc1=cmdscale(dist1,k=2)
> loc2=cmdscale(dist2,k=2)
> x1=loc1[,1]
> y1=loc1[,2]
> x2=loc2[,1]
> y2=loc2[,2]
> plot(x1,y1,type="n")
```

> *text(x1,y1,labels=creditdata_noout_noimp_train[,22],*

　　　　　　　*col=as.numeric(creditdata_noout_noimp_train[,22])+4)*

> *plot(x2,y2,type="n")*

> *text(x2,y2,labels=creditdata_noout_noimp_train_smot[,22],*

　　　　　　*col=as.numeric(creditdata_noout_noimp_train_smot[,22])+4)*

## Features Selection

*1)*　*Correlation Analysis:   Datasets may contain irrelevant or redundant features which might make the model more complicated. Hence removing such redundant features will speed up the model. The function plotcorr() plots a correlation matrix using ellipse shaped glyphs for each entry. It shows the correlation between the features in an easy way. The plot is coloured for more clarity. The following code displays the correlation. Correlation is checked independently for each data type: numeric and nominal. From the results in the figures, Fig. 7.8 and Fig. 7.9, it is observed that there is no positive correlation between any of the features, both numeric and quantitative. Hence, in this step none of the features are removed.*

> *library(ellipse)*

> *c= c(2,5,8,11,13,16,18)*

> *plotcorr(cor(creditdata_noout_noimp_train[,c]),col=cl<-c(7,6,3))*

> *c= c(1,3,4,6,7,9,10,12,14,15,17)*

> *plotcorr(cor(creditdata_noout_noimp_train [,c]),col=cl<-c("green","red","blue"))*

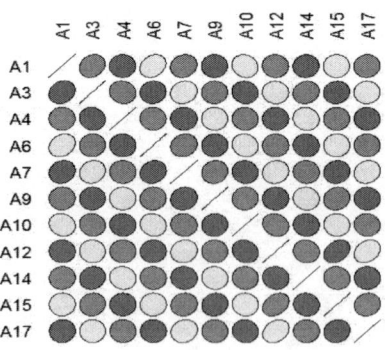

**Figure 7.8**   Correlation of Numeric Features

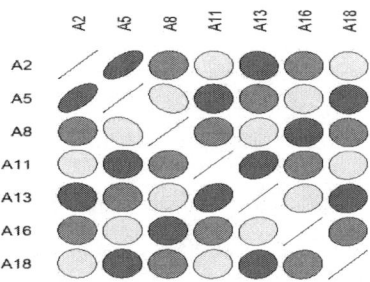

**Figure 7.9**   Correlation of Quantitative Features

2)   *Ranking Features:   The aim of this step is to find the subset of features that will be really relevant for the analysis as irrelevant features causes drawbacks like increased runtime, complex patterns etc. This resultant subset of features should give the same results as that of the original dataset. The proposed method picks a random object from the observations and generates several trees and on the basis of the accuracy of classifier or error ratio, features are weighted. To make the table of important features the following code is used.*

> *library(randomForest)*

> *set.seed(454)*

> *data.frame(creditdata_noout_noimp_train)*

> *randf<-randomForest(Def~ ., data=creditdata_noout_noimp_train, ntree=700,*

                         *importance=TRUE, proximity=TRUE)*

> *importance(randf, type=1, scale=TRUE)*

The above function *importance()* displays the features importance using the *"mean decrease accuracy"* measure in the below table. The measures can be plotted using the function *varImpPlot()* as shown in the figure. *Fig. 7.10.*

| Features | Mean Decrease Accuracy |
|----------|------------------------|
| A1       | 8.085083               |
| A2       | 7.070556               |
| A3       | 4.691744               |
| A4       | -0.10716               |

| Features | Mean Decrease Accuracy |
|----------|------------------------|
| A5 | 6.238347 |
| A6 | 4.554283 |
| A7 | 3.316346 |
| A8 | 0.59622 |
| A9 | 1.634721 |
| A10 | 1.383725 |
| A11 | 0.541585 |
| A12 | 2.344433 |
| A13 | 2.621854 |
| A14 | 4.629331 |
| A15 | 0.825801 |
| A16 | 1.225997 |
| A17 | 0.635881 |
| A18 | 0.037408 |
| A19 | 1.117891 |
| A20 | 1.388876 |

> *varImpPlot(randf)*

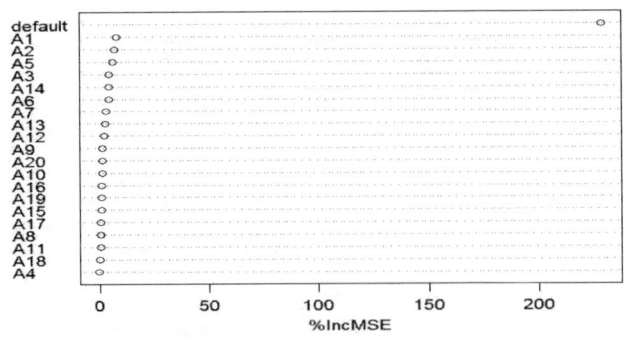

**Figure 7.10**    Plot Showing Importance of Features

3)  *Features Selection:   To fix the number of features to be selected based on the ranking, a threshold is required. This is accomplished using the below code.*

```
> findopt=rfcv(creditdata_noout_noimp_train[,-21],
> creditdata_noout_noimp_train[,21], cv.fold=10, scale="log", step=0.9)
> opt <- which.max(findopt$error.cv)
> plot( findopt$n.var, findopt$error.cv, type= "h", main = "Importance",
            xlab="Number of Features", ylab = "Classifier Error Rate")
> axis(1, opt, paste("Threshold", opt, sep="\n"), col = "red", col.axis = "red")
```

The result of this code is shown in the figure *Fig. 7.11* and it shows the best number of features is 15. Hence we select the features A1, A2, A3, A5, A6, A7, A9, A10, A12, A13, A14, A16, A19, A20, Def to build the model.

**Building Model**

Classification is one of the data analysis forms that predicts categorical labels. We used the decision tree model to predict the probability of default. The following code uses the function *rpart()* and finds a model from the training dataset.

```
> library(rpart)
> c = c(4, 8, 11, 15, 17, 18, 22)
> trdata=data.frame(creditdata_noout_noimp_train[,-c])
> tree=rpart(trdata$Def~.,data=trdata,method="class")
> printcp(tree)
```

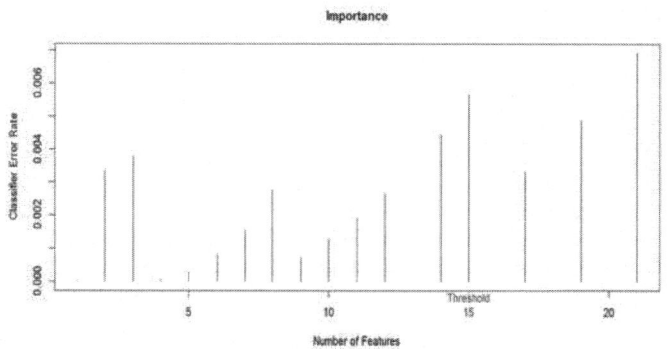

**Figure 7.11**   Threshold for Choosing the Number of Features

The result of this code is displayed below and in the table below.

*Classification tree:*

*rpart(formula = trdata$Def ~ ., data = trdata, method = "class")*

*Variables actually used in tree construction:*

*[1] A1  A12 A13 A2  A3  A5  A6  A9*

*Root node error: 232/780 = 0.29744*

*n= 780*

| CP | nsplit | rel | error | Xerror | xstd |
|----|--------|-----|-------|--------|------|
| 1 | 0.049569 | 0 | 1 | 1 | 0.05503 |
| 2 | 0.012931 | 4 | 0.78448 | 0.84483 | 0.052215 |
| 3 | 0.011494 | 5 | 0.77155 | 0.88793 | 0.053071 |
| 4 | 0.010057 | 9 | 0.72414 | 0.89655 | 0.053235 |
| 5 | 0.01 | 18 | 0.61207 | 0.89655 | 0.053235 |

The command to plot the classification tree is shown in the figure *Fig. 7.12.*

> *plot(tree, uniform=TRUE,main="Classification Tree")*

> *text(tree, use.n=TRUE, all=TRUE, cex=0.7)*

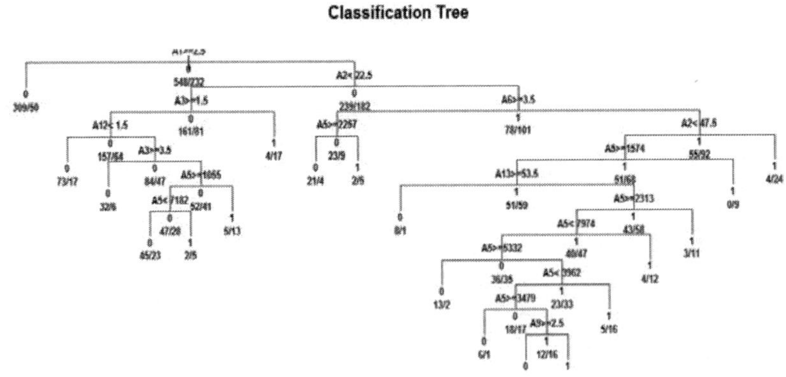

**Figure 7.12**   Classification Tree Model

## Prediction

The model is tested using the test dataset by using the *predict()* function. The code for the same and the results of the prediction are displayed below and in the table below.

> *predicttest=data.frame(testdata)*

> *pred=predict(tree,predicttest)*

> *c=c(21)*

> *table(predict(tree, testdata, type="class",na.action=na.pass), testdata[, c])*

|        | def | Nondef |
|--------|-----|--------|
| def    | 30  | 5      |
| nondef | 6   | 154    |

## Evaluation

Common metrics calculated from the confusion matrix are Precision, Accuracy, True Positive Rate (TP Rate) and False Positive Rate (FP Rate). The calculations for the same are listed below.

$$Precision = \frac{True\,Defaults}{True\,Defaults + False\,Defaults}$$

$$Accuracy = \frac{True\,Defaults + True\,Non\,defaults}{Total\,Test\,set}$$

$$TP\,Rate = \frac{True\,Defaults}{Total\,Defaults}$$

$$FP\,Rate = \frac{False\,Defaults}{Total\,Non\,defaults}$$

From our resultant data we get the values of the above metrics by applying the values as derived below.

| | | | | |
|---|---|---|---|---|
| *True Defaults* | = | 30 | | |
| *False Default* | = | 6 | | |
| *Total Default* | = | 35 | | |
| *True Non default* | = | 154 | | |
| *False Non default* | = | 5 | | |
| *Total Non default* | = | 160 | | |
| *Total Test set* | = | 195 | | |
| *Precision* | = | 30 / (30 + 6) | = | 0.833 |
| *Accuracy* | = | (30 + 154) / 195 | = | 0.943 |
| *TP Rate* | = | 30 / 35 | = | 0.857 |
| *FP Rate* | = | 6 / 160 | = | 0.037 |

| Precision | Accuracy | TP Rate | FP Rate |
|-----------|----------|---------|---------|
| 0.833 | 0.943 | 0.857 | 0.037 |

These results show that the proposed model is performing with high accuracy and precision and hence can be applied for credit scoring.

## 7.3. Exploratory Data Analysis on Crime Data

This case study is to do exploratory data analysis by inspecting, cleansing, transforming and find meaningful insights of the publicly available crime data of US. The goal is to discover useful information, suggest conclusion and support decision making. The dataset is loaded into the R tool and understood completely. The unwanted data are removed, that is the data is being pre-processed. As outcome the crime which leads to arrest, frequency of different crime types, when crime takes place are being found out. The result that is being found out is being visualized using bar plots and heat maps. As the results are visualized it is easy to draw conclusions. The objective is to identify criminal hot spots dynamically, to facilitate precautionary deployment of police resources, to perform temporal data analysis of crime data and to capture the trend of crimes happening.

This case study will have the below functions:

1. Gather and load the US Crime data into R Tool
   a. Understand the data field organization and its important dimensions
   b. Each crime incident indicates one record
   c. Find the different crime types
2. Pre-process the data
   a. Removing duplicate records
   b. Removing records with missing values
   c. Removing records with incorrect values
   d. Formatting the Timestamp field (splitting the date and time parts of the data)
   e. Binning the time intervals (4 intervals – each 6 hours)
   f. Grouping of similar crimes into one crime type
3. Exploratory analysis of the real time crime data
   a. Find and visualize which crime types lead to arrests
   b. Find and visualize frequency of different crime types
   c. Find and visualize the hours of the day in which more crimes happen
   d. Find and visualize the days of a week in which more crimes happen
   e. Find and visualize the months of a year in which more crimes occur
   f. Visualize which crime type happens more during which hours of the day
   g. Visualize which crime type happens more during which day of the week
   h. Visualize which crime type happens more during which month of the year

The packages used in this case study are:

*chron* - Creates chronological objects which represent dates and time of a day.

*ggplot2* - A system for creating graphs, based on the data you provide.

The function used are:

*read.csv()* - Reads a file in table format and creates a data frame from it.

*subset()* - Return subsets of vectors, matrices or data frames which meet conditions.

*is.na()* - NA is a logical constant of length 1 which contains a missing value indicator.

*which()* - Gives the TRUE indices of a logical object, allowing for array indices

*as.POSIXlt()* - Function to manipulate objects of class "POSIXlt" representing calendar dates and times.

*times()* - Creates objects which represent date or time.

*format()* - Format an R object for pretty printing.

*cut()* - Cut divides the range of x into intervals and codes the values in x according to which interval they fall. The leftmost interval corresponds to level one, the next leftmost to level two and so on.

*labels()* - Finds a suitable set of labels from an object for use in printing or plotting

*table()* - Table uses the cross-classifying factors to build a contingency table of the counts at each combination of factor levels.

*strptime()* - Function to convert between character representations and objects of class "POSIXlt" representing calendar dates and times.

*weekdays()* - Extracts the day of the week

*months()* - Extracts the month

*length()* - Get or set the length of vectors (including lists) and factors, and of any other R object for which a method has been defined.

*unique()* - Removes duplicated elements/rows from a vector, data frame or array

*as.character()* - Creates or tests for objects of type "character"

*ifelse()* - Ifelse returns a value with the same shape as test which is filled with elements selected from either yes or no depending on whether the element of test is TRUE or FALSE.

*factor()* - The function factor is used to encode a vector as a factor (the terms 'category' and 'enumerated type' are also used for factors). If argument ordered is TRUE, the factor levels are assumed to be ordered.

*aggregate()* - Splits the data into subsets, computes summary statistics for each, and returns the result in a convenient form.

*qplot()* - It's is the basic plotting function. It is used to create a different number of plots.

*ggplot()* - It initializes a ggplot object. It can be used to declare input data frame for a graph and to specify the set of plot aesthetics.

The dataset was analyzed to get details such as the file size, number of records, the fields specified and their meaning.

| Dataset Name | Size | No. of Attributes | No. of Records | Attribute Names |
|---|---|---|---|---|
| crime.csv | 44.9 MB | 17 | 262601 | Case, Date of occurrence, Block, IUCR, Primary description, Secondary description, Location description, Arrest, Domestic, Beat, Ward, FBI.CD, X coordinate, Y coordinate, Latitude, Longitude, Location |

U.S crime dataset is loaded into the R tool and the data field organization and its important dimensions are understood. It is noted that each crime incident indicates one record and the various crime types are manually analyzed. The dataset is loaded using the *read.csv()* function. After this the required packages for this project are installed and loaded using the below commands.

> *install.packages("chron")*

> *library(chron)*

> *install.packages("Rcpp", dependencies = TRUE)*

> *library(Rcpp)*

> *install.packages("ggplot2")*

> *library(ggplot2)*

The pre-processing deals with removing duplicate records, removing records with missing values, removing records with incorrect values, formatting the Timestamp field (splitting the date and time parts of the data), binning the time intervals (4 intervals – each 6 hours) and grouping of similar crimes into one crime type. The functions used for these preprocessing steps are *subset()*, *as.POSIXlt()*, *weekdays()*, *months()*, *chron()*, *cut()*, *table()*, *length()*, *as.character()*, *ifelse()*.

Then finally we go for finding and visualizing which crime types lead to arrests, finding and visualizing frequency of different crime types, finding and visualizing the hours of the day in which more crimes happen, finding and visualizing the days of a week in which more crimes happen, finding and visualizing the months of a year in which more crimes occurs, visualizing the occurrence of various crime types during various hours of the day, visualizing the occurrence of various crime types during various days of the week, visualizing the occurrence of various crime types during various months of the year. All these exploration and visualization are done using the functions *qplot()*, *factor()*, *aggregate()* and *ggplot()* that belong to the package *ggplot2*.

The code for the entire case study is as follows.

> *setwd("F:/")*

**#Loading data**

> *crime.data  <- read.csv("crime.data.csv")*

*#Removing duplicate records*

```
> crime.data  <- subset(crime.data, !duplicated(crime.data$CASE.))

> crime.data  <- subset(crime.data, !is.na(crime.data$LATITUDE))

> crime.data  <- subset(crime.data, !is.na(crime.data$WARD))
```

*#Removing incorrect records*

```
> crime.data  <- crime.data[crime.data$CASE.!=  "CASE#",]
```

*#Date Conversion*

```
> crime.data$date  <- as.POSIXlt(crime.data$DATE..OF.OCCURRENCE,
                                 format= "%m/%d/%Y %I:%M:%S %p")

> crime.data  <- subset(crime.data, !is.na(crime.data$date))

> crime.data$time <- times(format(crime.data$date, "%H:%M:%S"))
```

*#Binning Time Intervals*

```
> time.tag <- chron(times = c("00:00:00", "06:00:00", "12:00:00", "18:00:00",
                              "23:59:00"))

> crime.data$time.tag <- cut(crime.data$time, breaks = time.tag,
        labels = c("00-06","06-12", "12-18", "18-00"), include.lowest = TRUE)

> table(crime.data$time.tag)
```

*#Formating Date*

```
> crime.data$date <- as.POSIXlt(strptime(crime.data$date, format = "%Y-%m-%d"))

> crime.data$day <- weekdays(crime.data$date, abbreviate = TRUE)

> crime.data$month <- months(crime.data$date, abbreviate = TRUE)
```

*#Finding unique crime types*

```
> table(crime.data$PRIMARY.DESCRIPTION)

> length(unique(crime.data$PRIMARY.DESCRIPTION))

> crime.data$crime <- as.character(crime.data$PRIMARY.DESCRIPTION)
```

*#Grouping crime types*

```
> crime.data$crime <- ifelse(crime.data$crime %in% c("CRIM SEXUAL ASSAULT",
        "PROSTITUTION", "SEX OFFENSE"), "SEX", crime.data$crime)
```

```
> crime.data$crime <- ifelse(crime.data$crime %in% c("MOTOR VEHICLE
                    THEFT"), "MVT", crime.data$crime)
```

```
>      crime.data$crime       <-      ifelse(crime.data$crime      %in%
c("GAMBLING",INTERFERE   WITH  PUBLIC  OFFICER",  "INTERFERENCE
WITH PUBLIC          OFFICER",  "INTIMIDATION",  "LIQUOR  LAW
VIOLATION", "OBSCENITY", "NON CRIMINAL", "PUBLIC PEACE VIOLATION",
"PUBLIC   INDECENCY",   "STALKING",   "NON-CRIMINAL  (SUBJECT
SPECIFIED)"), "NONVIO", crime.data$crime)
```

```
> crime.data$crime <- ifelse(crime.data$crime == "CRIMINAL DAMAGE",
                    "DAMAGE", crime.data$crime)
```

```
> crime.data$crime <- ifelse(crime.data$crime == "CRIMINAL
            TRESPASS","TRESPASS", crime.data$crime)
```

```
> crime.data$crime <- ifelse(crime.data$crime %in% c("NARCOTICS",
"OTHERNARCOTIC VIOLATION", "OTHER NARCOTIC VIOLATION"),
                    "DRUG", crime.data$crime)
```

```
> crime.data$crime <- ifelse(crime.data$crime == "DECEPTIVE
            PRACTICE","FRAUD", crime.data$crime)
```

```
> crime.data$crime <- ifelse(crime.data$crime %in% c("OTHER OFFENSE",
            "OTHEROFFENSE"), "OTHER", crime.data$crime)
```

```
> crime.data$crime <- ifelse(crime.data$crime %in% c("KIDNAPPING",
    "WEAPONSVIOLATION", "OFFENSE INVOLVING CHILDREN"), "VIO",
                    crime.data$crime)
```

```
> table(crime.data$crime)
```

```
> length(unique(crime.data$crime))
```

*#Finding crimes that leads to arrest*

```
> crime.data$ARREST <- ifelse(as.character(crime.data$ARREST) == "Y", 1, 0)
```

```
> crime.data.arrest <- crime.data[crime.data$ARREST != 0,]
```

> *table(crime.data.arrest$crime)*

*#Visualizing Crimes that leads to arrest*

> *qplot(crime.data.arrest$crime, xlab = "Crime", main = "Crimes in Chicago that*
> *leads to*

*arrest") + scale_y_continuous("Number of crimes") + theme(text = element_*
*text(size=10),*

*axis.text.x = element_text(angle=90, hjust=1))*

*#Visualizing Various Crime Types*

> *qplot(crime.data$crime, xlab = "Crime", main = "Crimes in Chicago") +*
>   *scale_y_continuous("Number of crimes") + theme(text = element_text(size=10),*
>                       *axis.text.x = element_text(angle=90, hjust=1))*

*#Visualizing Various Crimes by time of day*

> *qplot(crime.data$time.tag, xlab="Time of day", main = "Crimes by time of day") +*
>                       *scale_y_continuous("Number of crimes"*

*#Visualizing Various Crimes by day of week*

> *crime.data$day <- factor(crime.data$day,*
>             *levels = c("Mon", "Tue", "Wed", "Thu", "Fri", "Sat", "Sun"))*
> *qplot(crime.data$day, xlab = "Day of week", main= "Crimes by day of week") +*
>                   *scale_y_continuous("Number of crimes")*

*#Visualizing Various Crimes by month*

> *crime.data$month <- factor(crime.data$month, levels = c("Jan", "Feb", "Mar",*
>             *"Apr", "May", "Jun", "Jul", "Aug", "Sep", "Oct", "Nov", "Dec"))*
> *qplot(crime.data$month, xlab = "Month", main= "Crimes by month") +*
>                   *scale_y_continuous("Number of crimes")*

*#Visualizing Crimes by Time of Day using Heatmap*

> *temp <- aggregate(crime.data$crime, by= list(crime.data$crime,crime.data$time.*
>                   *tag), FUN= length)*

```
> names(temp) <- c("crime", "time.tag", "count")
> ggplot(temp, aes(x = crime, y = factor(time.tag))) + geom_tile(aes(fill = count)) +
                    scale_x_discrete("Crime", expand = c(0,0)) +
                 scale_y_discrete("Time of day", expand = c(0,-2)) +
scale_fill_gradient("Number of crimes", low = "white", high = "steelblue") +
                    theme_bw() + ggtitle("Crimes by time of day") +
               theme(panel.grid.major = element_line(colour = NA),
                  panel.grid.minor = element_line (colour = NA)) +
theme(text = element_text(size=10), axis.text.x = element_text(angle=90, hjust=1))
```

### #Visualizing Crimes by Day of Week using Heatmap

```
> temp <- aggregate(CASE. ~ crime + day, data = crime.data, FUN = length)
> names(temp)[3] <- "count"
> ggplot(temp, aes(x = crime, y = day, fill = count)) + geom_tile(aes(fill = count)) +
                    scale_x_discrete("Crime", expand = c(0,0)) +
                 scale_y_discrete("Day of week", expand = c(0,-2)) +
    scale_fill_gradient("Number of crimes", low = "white", high = "steelblue") +
                    theme_bw() + ggtitle("Crimes by day of week") +
               theme(panel.grid.major = element_line(colour = NA),
                  panel.grid.minor = element_line(colour = NA)) +
theme(text = element_text(size=10), axis.text.x = element_text(angle=90, hjust=1))
```

### #Visualizing Crimes by Month of Year using Heatmap

```
> temp <- aggregate(CASE. ~ crime + month, data = crime.data, FUN = length)
> names(temp)[3] <- "count"
> ggplot(temp, aes(x = crime, y = month, fill = count)) + geom_tile(aes(fill = count)) +
                    scale_x_discrete("Crime", expand = c(0,0)) +
                 scale_y_discrete("Month", expand = c(0,-2)) +
```

*scale_fill_gradient("Number of crimes", low = "white", high = "steelblue") +*

*theme_bw() + ggtitle("Crimes by Month") +*

*theme(panel.grid.major = element_line (colour = NA),*

*panel.grid.minor = element_line(colour = NA)) +*

*theme(text = element_text(size=10), axis.text.x = element_text(angle=90, hjust=1)) +*

*theme(text = element_text(size=10), axis.text.x = element_text(angle=90, hjust=1))*

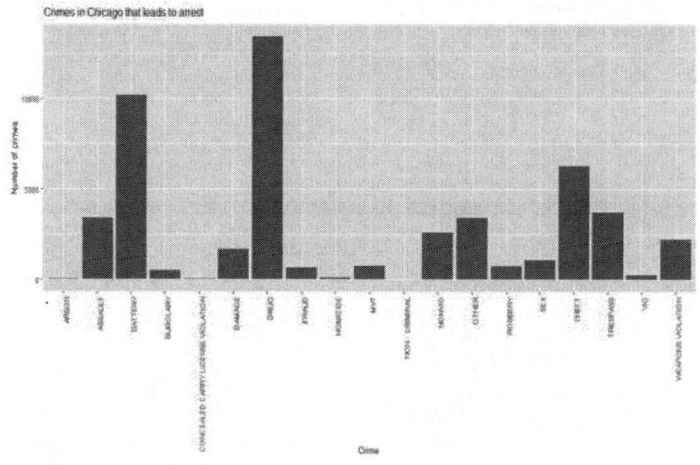

**Figure 7.13**   Ggplot of Crimes Leading to Arrest

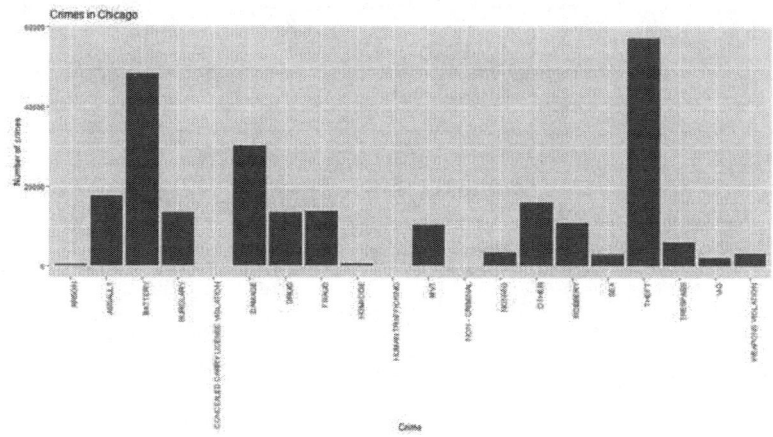

**Figure 7.14**   Frequencies of Crimes

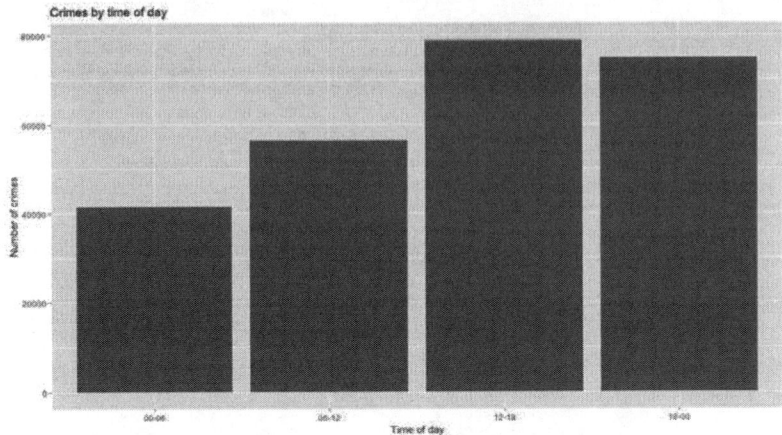

**Figure 7.15** Ggplot of Crimes by Time of the Day

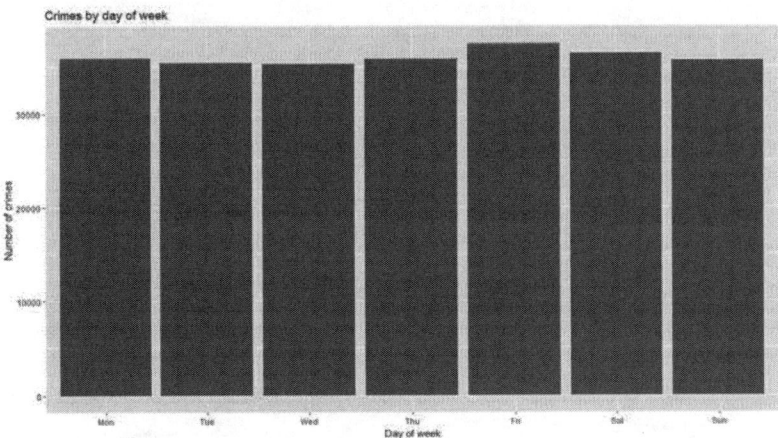

**Figure 7.16** Ggplot of Crimes by Day of the Week

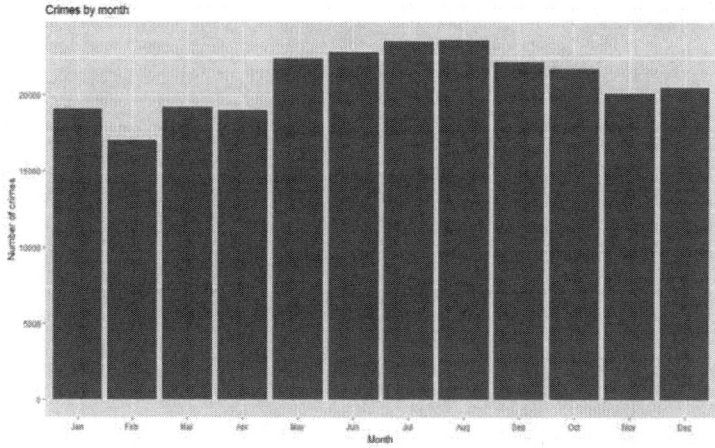

**Figure 7.17**  Ggplot of Crimes by Month

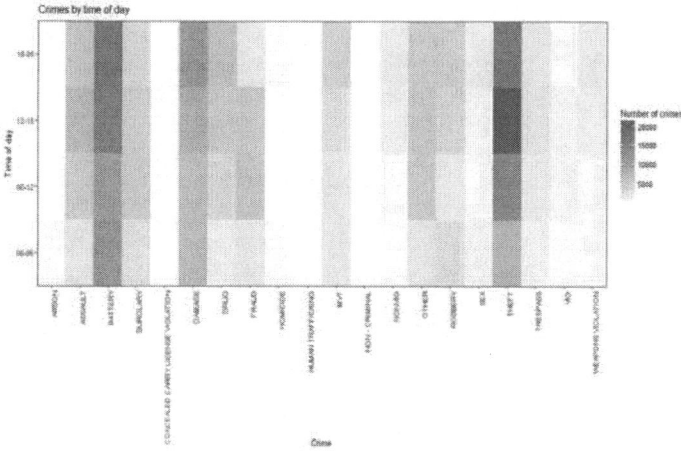

**Figure 7.18**  Heatmap of Crimes by Time of the Day

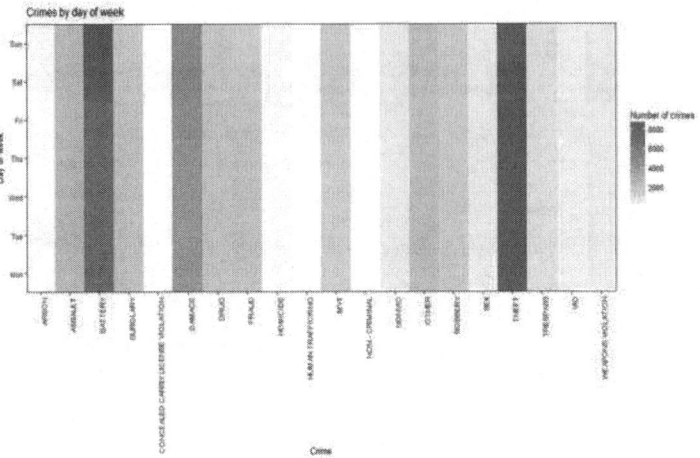

**Figure 7.19**   Heatmap of Crimes by Day of the Week

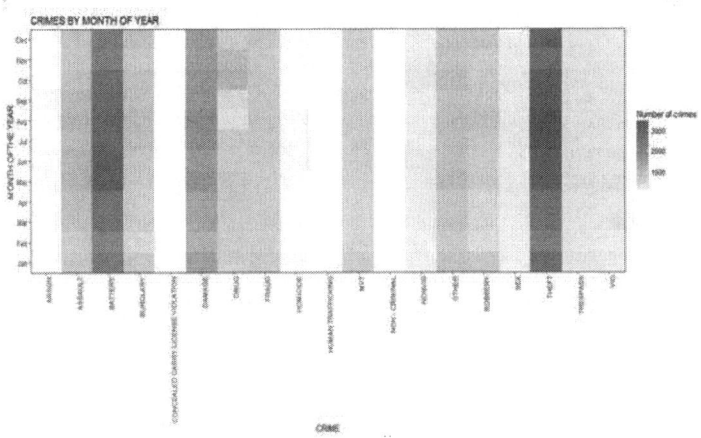

**Figure 7.20**   Heatmap of Crimes by Month of the Year

## 7.4. Exploratory Data Analysis on Baseball Data

This case study is to find meaningful information by visualization about the Base Ball Teams by using the packages and functions in R. Data is spread across three files, namely, salaries.csv, master.csv and batting.csv. These datasets are loaded and sequences of pre-processing are done. The pre-processed data is then analyzed on several aspects such as variation of salary over a period of time with respect to League, Team etc. The player's career records are analyzed based on their hits and runs and their batting averages calculated. These analysis results are then presented in the form of graphs and histograms using the functions available in R. The objective is to identify the trend of base ball players salary over the years, facilitates to understand the correlation between players salary and their performances, analyze if age, country, height and weight of the players have impact on their performance and it captures the details of top performing baseball players.

This case study will have the below functions:

1. Gather and load the US Baseball data into R Tool
   a. Understand the data field organization and its important dimensions
   b. The data is split among three files namely Master.csv, Batting.csv and Salaries.csv

2. Pre-process the data
   a. Filtering records specific to a particular periods and specific leagues
   b. Sorting the filtered data based on the chronological order of records
   c. Summarising data to get min, max and average salary year wise
   d. Summarising data to get min, max and average salary league wise
   e. Summarising data to get min, max and average salary team wise
   f. Summarising to get min, max and average salary year wise and league wise
   g. Summarising to get min, max and average salary year wise and team wise
   h. Merging the salary data of the base ball players with their biographical (personal) data – (To combine player's last and first name into single name)
   i. Merging the salary data with batting statistics (history data) data based on the year, player, Team and League fields (To study salary given based on performance)
   j. Merging the salary, master and batting datasets (To study player wise batting performance)

     k.    Filter the records of the players in the years in which they have not had a chance to bat (AB > 0)

  c.  Exploratory analysis of the baseball team data

     a.    Visualize the trend of how salaries change over time

     b.    Find one players salary, team and other details

     c.    Find the relation of the player's salary with his height, weight, birth country

     d.    Find how each player was batting year wise

     e.    Visualize correlation of salary with the players performance

     f.    Visualize each player's career record (Eg. Total Hit and Runs) based on their highest rank

     g.    Visualize the correlation between the players hits and runs

     h.    Visualize the batting average of the players in a histogram

The packages used in this case study are:

**data.table** - Extension of package "data.frame". Does fast aggregation of large data, ordered joins, add/modify/delete of columns by group, a fast friendly file reader and parallel file writer.

**ggplot2** - A system for creating graphs, based on the data you provide.

The function used are:

**read.csv()** - Reads a file in table format and creates a data frame from it.

**as.data.table()** - It is a function to check if an object is of type "data.table".

**merge()** - It is used to merge two data frames by common column or row names, or do other versions of database join operations.

**order()** - Order returns a permutation which rearranges its first argument into ascending or descending order, breaking ties by further arguments.

**max()** - Returns the (parallel) maxima and minima of the input values.

**min()** - Returns the (parallel) maxima and minima of the input values.

*mean()* - Generic function for the (trimmed) arithmetic mean.

*paste()* - Paste converts its arguments to character strings, and concatenates them. If the arguments are vectors, they are concatenated term-by-term to give a character vector result.

*ggplot()* - initializes a ggplot object. It can be used to declare the input data frame for a graphic and to specify the set of plot aesthetics intended to be common throughout all subsequent layers unless specifically overridden.

The dataset was analyzed to get details such as the file size, number of records, the fields specified and their meaning.

| Dataset Name | Size | No. of Attributes | No. of Records | Attribute Names |
| --- | --- | --- | --- | --- |
| salaries.csv | 707 KB | 5 | 23956 | yearID, teamID, lgID, playerID, salary |
| master.csv | 2.31 MB | 24 | 18355 | playerID, birthYear, birthMonth, birthDay, birthCountry, birthState, birthCity, deathYear, deathMonth, deathDay, deathCountry, deathState, deathCity, nameFirst, nameLast, nameGiven, weight, height, bats, throws, debut, finalGame, retroID, bbrefID |
| batting.csv | 6.27 MB | 24 | 97890 | playerID, yearID, stint, teamID, lgID, G, G_batting, AB, R, H, 2B, 3B, HR, RBI, SB, CS, BB, SO, IBB, HBP, SH, SF, GIDP, G_old |

US Baseball team datasets *salaries.csv*, *master.csv* and *batting.csv* are loaded into the R tool and the data field organization and its important dimensions are understood. It is noted that each record in the master has a corresponding record in the salaries and batting datasets. The relationships between these datasets were manually analyzed. The datasets are loaded using the *read.csv()* function of the R Package. After this the required packages for this project are installed and loaded using the below commands.

> *install.packages("data.table")*

> *library(data.table)*

> *install.packages("Rcpp", dependencies = TRUE)*

> *library(Rcpp)*

> *install.packages("ggplot2")*

> *library(ggplot2)*

Data preprocessing describes any type of processing performed on raw data to prepare it for further analysis. In this step, datasets are processed using *data.table* package and their functions. These pre-processing steps are very essential to get accuracy.

The Pre-Processing steps are:

(i) **Filtering**

In this process, the salaries dataset is filtered based on their fields in different dimensions such as filtering records specific to a particular periods and specific leagues (NL or AL), filtering the records of the players in the years in which they have not had a chance to bat (AB > 0).

(ii) **Sorting**

Sorting is the process for ordering the data in a correct format. Here the filtered data should be sorted based on the chronological order of records

(iii) **Summarizing**

This deals with summarising data to get minimum, maximum and average salary year wise, team wise and league wise.

(iv) **Merging**

This step deal with merging the salary data of the base ball players with their biographical (personal) data (To combine player's last and first name into single name), merging the salary data with batting statistics (history data) data based on the year, player, Team and League fields (To study salary given based on performance), merging the salary, master and batting datasets (To study player wise batting performance), filtering the records of the players in the years in which they have not had a chance to bat (AB > 0) and merging the three datasets to get a study of player wise batting performance.

Exploration and Visualization deals with finding the trend of Salaries of Players over the Years (Entire Data Set), finding trend of Salaries of Players over the Years >= 1990 and only for the American League, finding the year wise Average Salary, finding Year wise & League wise Average Salary, finding Year wise & Team wise Average Salary, finding the correlation between the players Hits and Runs, finding one Players Salary details in the different years and the teams he belonged to, finding Batting Average of the Players, visualizing the graph of Players Salary Vs Height, Players Salary Vs Weight and Players Salary Vs Birth Country. All these exploration and visualization are done using the function *ggplot()* that belongs to the package *ggplot2*.

The code for the entire case study is as follows.

> *setwd("F:")*

### #Loading Data

> *salaries<-read.csv("salaries.csv",header=TRUE)*

> *master<-read.csv("master.csv",header=TRUE)*

> *batting<-read.csv("batting.csv",header=TRUE)*

### #Transforming into data table

> *salaries=as.data.table(salaries)*

> *batting=as.data.table(batting)*

> *master=as.data.table(master)*

### #Data Filtered, sorted and grouped

> *salaries.filtered=salaries[lgID == "AL" & yearID >= 1990, ]*

> *salaries.filtered.sorted=salaries.filtered[order(salary), ]*

> *summarized.year=salaries[ ,list(Average=mean(salary)), by="yearID"]*

```
> summarized.lg=salaries[ ,list(Average=mean(salary),Maximum
                          =max(salary),minimum=min(salary)),by="lgID"]
> summarized.year.lg=salaries[,list(Average=mean(salary),Maximum=max(salary),
                          minimum=min(salary)),by=c("yearID","lgID")]
> summarized.year.play=salaries[,list(Average=mean(salary),Maximum=max(salary),
                          minimum=min(salary)),by=c("yearID","playerID")]
> summarized.year.team=salaries[,list(Average=mean(salary),Maximum=max(salary),
                          minimum=min(salary)),by=c("yearID","teamID")]
> batting=as.data.table(batting)
```

#### #Data merged and reduced

```
> merged.sm=merge(salaries,master,by="playerID")
> merged.sm$name<-paste(merged.sm$nameFirst,merged.sm$nameLast)
> merged.batting= merge(batting, salaries, by=c("playerID", "yearID", "teamID",
                          "lgID"), all.x=TRUE)
> merged.all=merge(merged.batting,master,by="playerID")
> merged.all = merged.all[AB > 0, ]
> merged.all[, name:=paste(nameFirst, nameLast)]
> salaries$salary.reduced<-paste(salaries$salary/10000)
> z <- salaries$salary.reduced
> salary.reduced.round.<-round(as.numeric(z,1))
```

#### #Visualizing Trend of Salaries

```
> ggplot(salaries, aes(x=yearID, y=salary.reduced.round.))+geom_point()+
                          ylab("Salary (in 10000 USD)")+xlab("Year")
> salaries.filtered$salary.reduced<-paste(salaries.filtered$salary/10000)
> y<-salaries.filtered$salary.reduced
```

> *salaries.reduced.round<-round(as.numeric(y,1))*

#### #Visualizing Trend of Salaries of Year > 1990 for American League

> *ggplot(salaries.filtered, aes(x=yearID, y=salaries.reduced.round))+geom_point()+*
> ylab("Salary (in 10000 USD)")+xlab("Year")

> *summarized.year$average.reduced<-paste(summarized.year$Average/10000)*

> *x<-summarized.year$average.reduced*

> *average.reduced.round<-round(as.numeric(x,1))*

#### #Visualizing Year wise Average Salary

> *ggplot(summarized.year, aes(x=yearID, y=average.reduced.round))+geom_line()+*

> ylab("Salary (in 10000 USD)")+xlab("Year")

> *summarized.year.lg$average.reduce.round<-paste(summarized.year.*
> *lg$Average/10000)*

> *averages.reduce.round<-round(as.numeric(summarized.year.lg$average.reduce.*
> *round,1))*

#### #Visualizing Year wise and League wise Average Salary

> *ggplot(summarized.year.lg, aes(x=yearID, y=averages.reduce.round, color=lgID))+*
> *geom_smooth(se=FALSE)+ylab ("Average (in 10000 USD)")+xlab("Year")*

> *summarized.year.team$average.reduce.round<-*
> *paste(summarized.year.team$Average/10000)*

> *averages.reduce.round <-*
> *round(as.numeric(summarized.year.team$average.reduce.round,1))*

#### #Visualizing Year wise and Team wise Average Salary

> *ggplot(summarized.year.team, aes(x=yearID, y=averages.reduce.round,*
> *color=teamID))+geom_path()+ylab("Average (in 10000 USD)")+xlab("Year")*

#### #Visualizing Correlation between the Players Hits and Runs

> *ggplot(merged.all, aes(x=H, y=R,color=Total.HR))+geom_point()*

```
> salaries[salaries$playerID == "aardsda01"]
> merged.all$batting.average<-paste(merged.all$AB/merged.all$H)
```

#### #Visualizing Batting Average of Players

```
> ggplot(merged.all, aes(x=batting.average))+geom_histogram(stat = "count")
> merged.all$salary.reduced<-paste(merged.all$salary/10000)
> salary.reduced.round<-round(as.numeric(merged.all$salary.reduced,1))
```

#### #Visualizing Players Salary Vs. Height

```
> ggplot(merged.all, aes(x=height.in.inches.,y=salary.reduced.round))+
                             geom_smooth(se=FALSE)+
ylab("Salary (in 10000 USD)")+xlab("Height (in inches)")
```

#### #Visualizing Players Salary Vs. Weight

```
> ggplot(merged.all, aes(x=weight.in.pound.,
                   y=salary.reduced.round,color=weight.in.pound.))+
        geom_point()+ylab("Salary (in 10000 USD)")+xlab("Weight (in pound)")
```

#### #Visualizing Players Salary Vs. Birth Country

```
> ggplot(merged.all, aes(x=birthCountry, y=salary.reduced.
                       round,color=birthCountry))+
               geom_point()+theme(text=element_text(size = 10),
               axis.text.x=element_text(angle=90,hjust=1))+
               ylab("Salary (in 10000 USD)")+xlab("Birth Country")
```

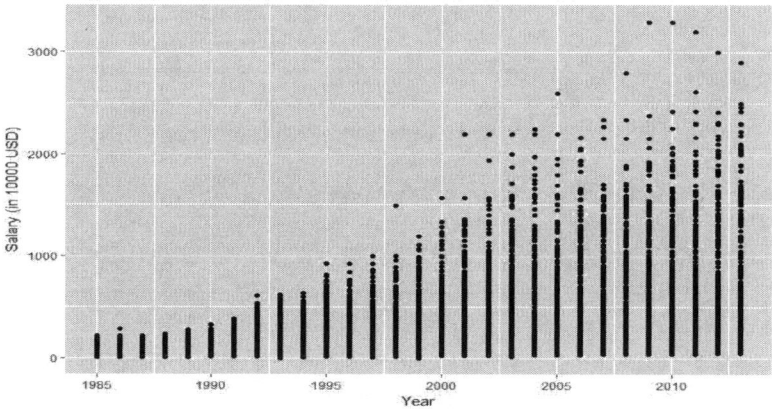

**Figure 7.21**   Ggplot of Trend of Salaries

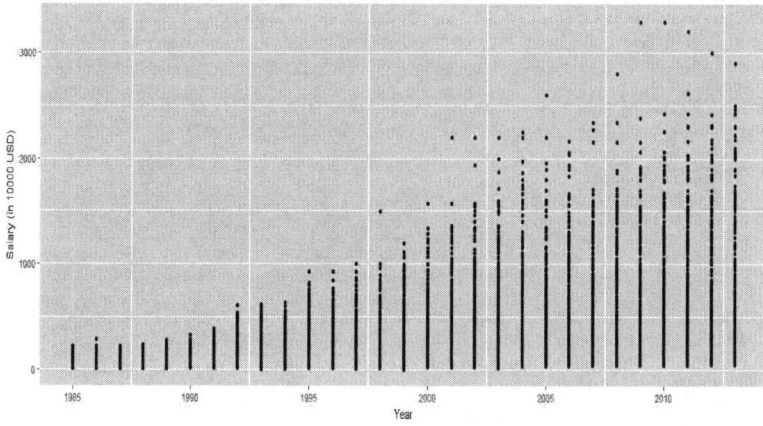

**Figure 7.22**   Ggplot of Trend of Salaries of Year > 1990 for American League

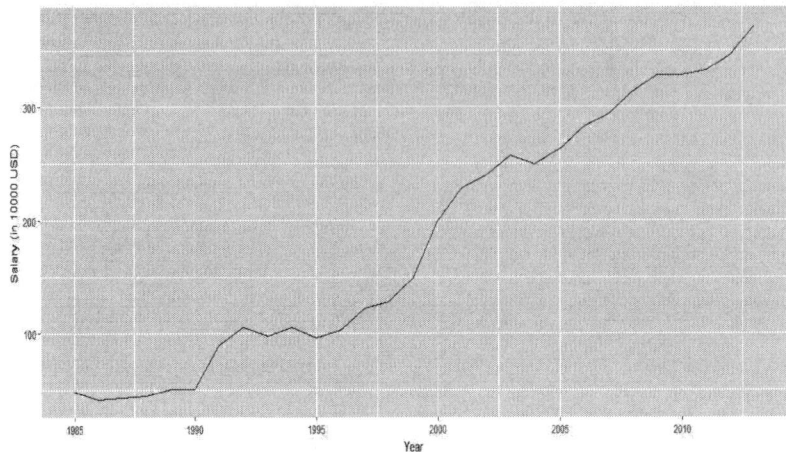

**Figure 7.23** Ggplot of Year wise Average Salary

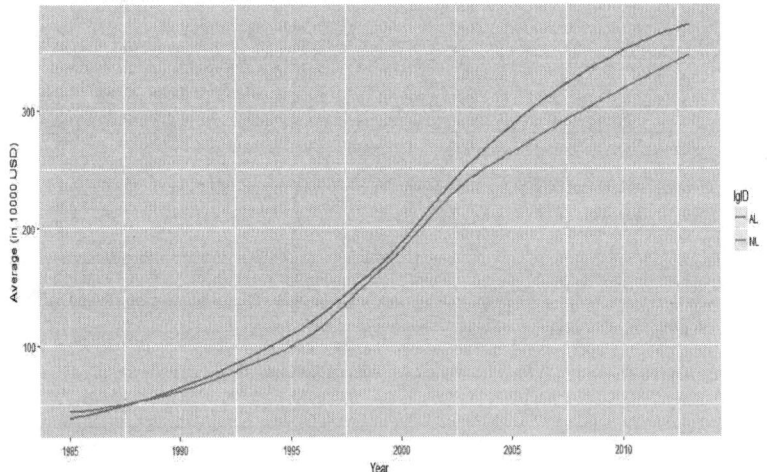

**Figure 7.24** Ggplot of Year wise and League wise Average Salary

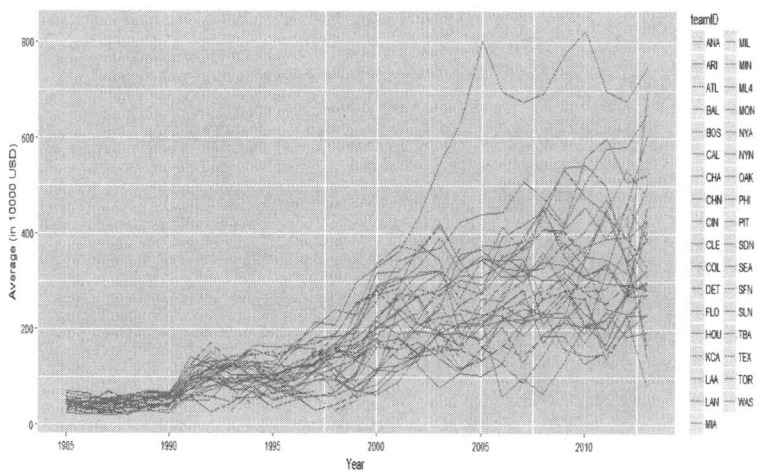

**Figure 7.25** Ggplot of Year wise and Team wise Average Salary

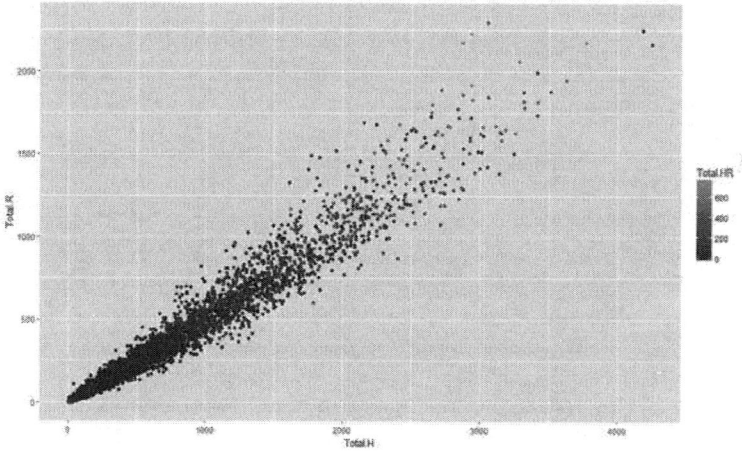

**Figure 7.26** Ggplot of Correlation between the Players Hits and Runs

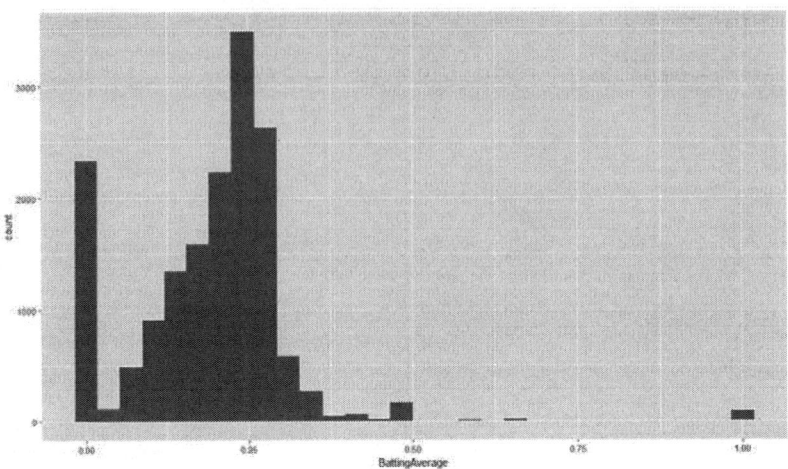

**Figure 7.27**   Ggplot of Batting Average of Players

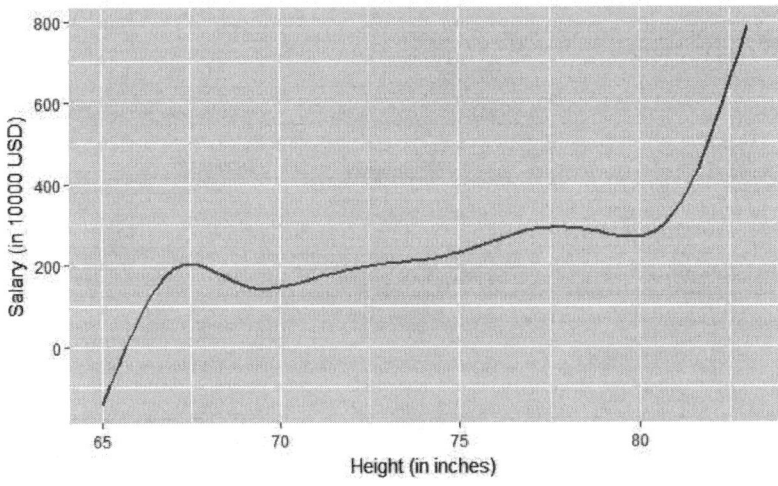

**Figure 7.28**   Ggplot of Players Salary Vs. Height

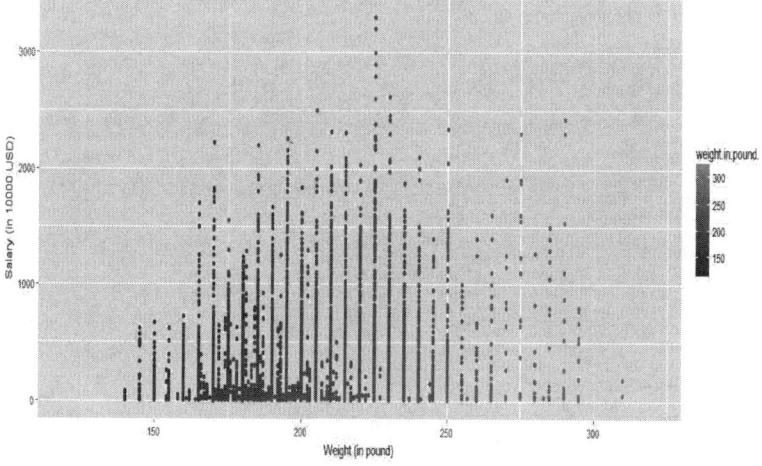

**Figure 7.29** Ggplot of Players Salary Vs. Weight

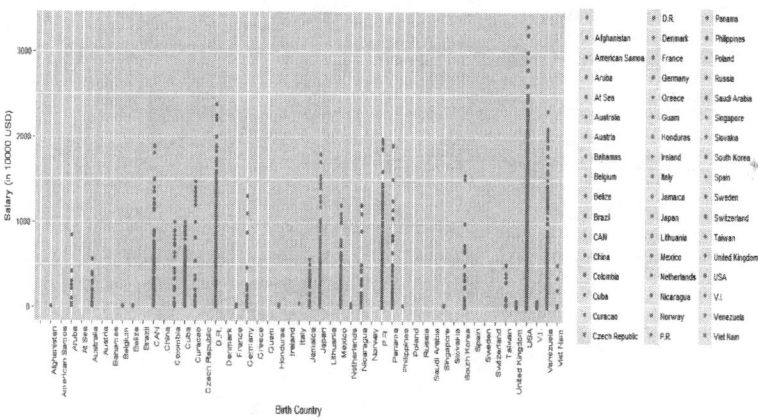

**Figure 7.30** Ggplot of Players Salary Vs. Birth Country

# 7.5. Social Network Analysis

Social Network Analysis (SNA) deals with finding social structures using the concepts of networks and graph theory. It characterizes networked structures in terms of nodes and the ties, edges, or links that connect them. Social Network Analysis in R is done using the package *igraph*. The data used for analysis is the twitter text data used in the *Text Mining* Section. The terms in this data can be considered as people and the tweets as groups in the *LinkedIn*. The *term-document matrix* is a representation of the group membership of people.

In this case study, we first build the network of terms based on their co-occurrence in the same tweets and then build a network of tweets based on the terms shared by them. After this, we also build a two-mode network composed of both terms and tweets.

As a first step we build the term-document matrix as in the *Text Mining* section using the below code.

```
> library(tm)
#Reading the input file
> dat<-read.csv("GameReview.csv",stringsAsFactors = FALSE)
#Converting it to a corpus
> corp <- Corpus(VectorSource(dat$text))
#Preprocessing – removing stop words, punctuations, whitespaces etc.
> corp <- tm_map(corp, content_transformer(tolower))
> corp <- tm_map(corp, removePunctuation)
> corp <- tm_map(corp, stripWhitespace)
> corp <- tm_map(corp, removeWords, stopwords("english"))
#Converting into a term-document matrix
> tdm <- TermDocumentMatrix(corp)
#Removing sparse terms
> tdm2 <- removeSparseTerms(tdm, sparse=0.96)
#converting into a matrix
> termDocMatrix <- as.matrix(tdm2)
> termDocMatrix <- termDocMatrix[ ,1:150]
```

After that the *term-document matrix* is transformed into a *term-term adjacency matrix*, based on which a graph is built. The graph is then plotted to show the relationship between the frequent terms and also make the graph more readable by setting colors, font sizes and transparency of vertices and edges.

> *termDocMatrix[5:10,1:20]*

    Docs

| Terms | 1 | 2 | 3 | 4 | 5 | 6 | 7 | 8 | 9 | 10 | 11 | 12 | 13 | 14 | 15 | 16 | 17 | 18 | 19 | 20 |
|-------|---|---|---|---|---|---|---|---|---|----|----|----|----|----|----|----|----|----|----|----|
| good  | 0 | 1 | 0 | 0 | 1 | 1 | 0 | 0 | 0 | 0  | 0  | 0  | 0  | 0  | 0  | 0  | 1  | 1  | 0  | 0  |
| great | 0 | 1 | 0 | 0 | 1 | 0 | 0 | 0 | 1 | 0  | 2  | 0  | 0  | 0  | 1  | 0  | 0  | 0  | 0  | 0  |
| money | 0 | 1 | 0 | 0 | 0 | 0 | 0 | 0 | 0 | 0  | 0  | 0  | 0  | 0  | 0  | 0  | 0  | 0  | 0  | 0  |
| play  | 0 | 1 | 0 | 0 | 0 | 2 | 0 | 0 | 0 | 0  | 0  | 1  | 0  | 0  | 0  | 0  | 0  | 0  | 0  | 0  |
| fun   | 0 | 0 | 2 | 0 | 0 | 0 | 0 | 0 | 0 | 0  | 0  | 1  | 0  | 0  | 0  | 2  | 0  | 0  | 0  | 0  |
| much  | 0 | 0 | 1 | 0 | 0 | 0 | 0 | 0 | 0 | 0  | 0  | 0  | 0  | 0  | 0  | 0  | 0  | 0  | 0  | 0  |

> *termDocMatrix[termDocMatrix>=1] <- 1*

> *termMatrix <- termDocMatrix %\*% t(termDocMatrix)*

> *termMatrix[5:10,5:10]*

    Terms

| Terms | good | great | money | play | fun | much |
|-------|------|-------|-------|------|-----|------|
| good  | 29   | 6     | 3     | 8    | 4   | 1    |
| great | 6    | 25    | 1     | 5    | 5   | 1    |
| money | 3    | 1     | 9     | 3    | 1   | 2    |
| play  | 8    | 5     | 3     | 25   | 10  | 2    |
| fun   | 4    | 5     | 1     | 10   | 35  | 4    |
| much  | 1    | 1     | 2     | 2    | 4   | 8    |

In the above code, %\*% is an operator for the product of two matrices, and the function *t()* transposes a matrix. We then build a *term-term adjacency matrix*. In this matrix the rows and columns represent terms, and every entry is the number of concurrences of two terms. Next we can build a graph with the function *graph. adjacency()* from package *igraph*.

> *library(igraph)*

```
> g <- graph.adjacency(termMatrix, weighted=T, mode="undirected")
> g <- simplify(g)
> V(g)$label <- V(g)$name
> V(g)$degree <- degree(g)
```

After that, we plot the network with the function *layout.fruchterman.reingold()* as in the *Fig. 7.31*.

```
> set.seed(3952)
> layout1 <- layout.fruchterman.reingold(g)
> plot(g, layout=layout1)
```

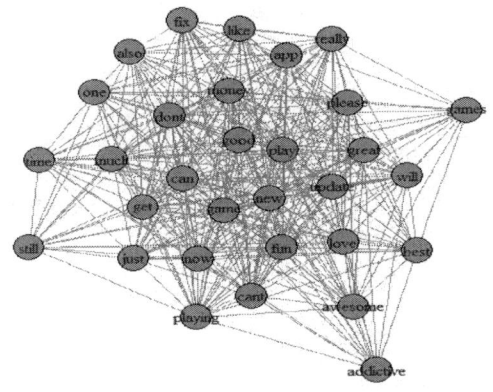

**Figure 7.31**  Network of Terms Based on their Co-occurrence

Next, we set the label size of vertices based on their degrees, to make important terms stand out. Similarly, we also set the width and transparency of edges based on their weights. This is useful in applications where graphs are crowded with many vertices and edges. The vertices and edges in the below code are accessed with V() and E(). The function *rgb(red, green, blue, alpha)* defines the colors. With the same layout, we plot the graph again as in *Fig. 7.32*.

```
> V(g)$label.cex <- 2.2 * V(g)$degree / max(V(g)$degree) + .2
> V(g)$label.color <- rgb(0, 0, .2, .8)
```

> V(g)$frame.color <- NA

> egam <- (log(E(g)$weight)+.4) / max(log(E(g)$weight)+.4)

> E(g)$color <- rgb(.5, .5, 0, egam)

> E(g)$width <- egam

> plot(g, layout=layout1)

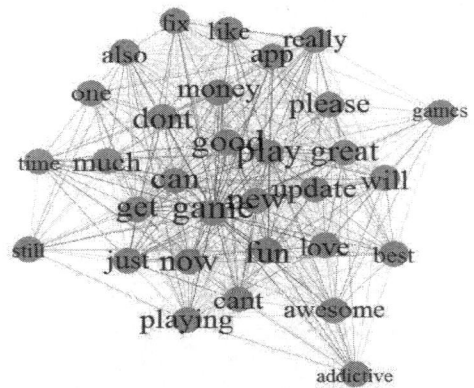

**Figure 7.32**   Network of Terms with Label Size of Vertices Based on their Degrees

Next, we try to detect communities from the graph, called as cohesive blocks and then plot the network of terms based on cohesive blocks as in *Fig. 7.33*.

> blocks <- cohesive.blocks(g)

> blocks

*Cohesive block structure:*

*B-1*     *c 15, n 31*

'- *B-2*     *c 16, n 30*   *oooooooooo .ooooooooo ooooooooooo o*

  '- *B-3*   *c 17, n 28*   *ooo.oooooo .ooooooooo ooooooooooo .*

> plot(blocks, g, vertex.size=.3, vertex.label.cex=1.5, edge.olor=rgb(.4,.4,0,.3))

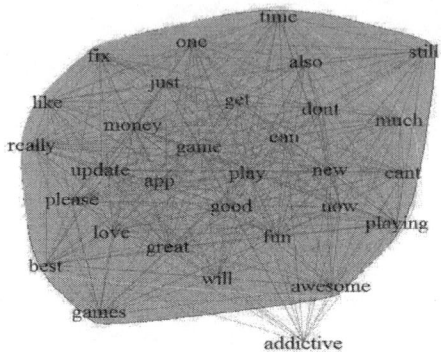

**Figure 7.33** Network of Terms with Communities called as Cohesive Blocks

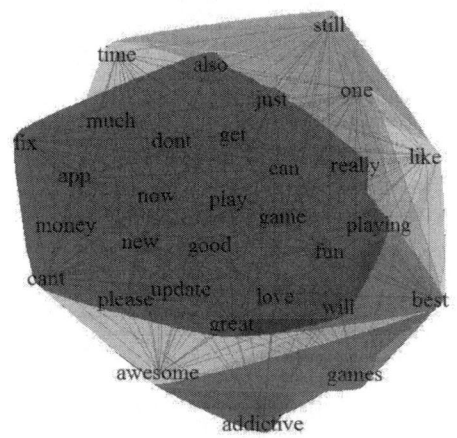

**Figure 7.34** Network of Terms Based on Maximal Cliques

Next we plot the network of terms based on maximal cliques as in *Fig. 7.34*.

> *cl <- maximal.cliques(g)*

> *length(cl)*

*[1] 286*

> *colbar <- rainbow(length(cl) + 1)*

```
> for (i in 1:length(cl)) {
+    V(g)[cl[[i]]]$color <- colbar[i+1]
+ }
> plot(g, mark.groups=cl, vertex.size=.3, vertex.label.cex=1.5, edge
                                         color=rgb(.4,.4,0,.3))
```

Next we plot the network of terms based on largest cliques as in *Fig. 7.35*.

```
> cl <- largest.cliques(g)
> length(cl)
[1] 41
> colbar <- rainbow(length(cl) + 1)
> for (i in 1:length(cl)) {
+    V(g)[cl[[i]]]$color <- colbar[i+1]
+ }
> plot(g, mark.groups=cl, vertex.size=.3, vertex.label.cex=1.5, edge.
                                         color=rgb(.4,.4,0,.3))
```

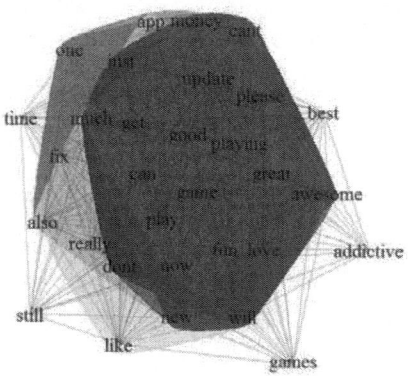

**Figure 7.35** Network of Terms Based on Largest Cliques

It is also possible to build a graph of tweets based on the number of terms that they have in common. Because most tweets contain the words *"game"*, *"play"*, most tweets are connected with others and the graph of tweets is very crowded. To simplify the graph and find relationship between tweets beyond the above two keywords, we remove the two words before building a graph.

> *idx <- which(dimnames(termDocMatrix)$Terms %in% c("game", "play"))*

> *M <- termDocMatrix[-idx,]*

> *tweetMatrix <- t(M) %\*% M*

> *g <- graph.adjacency(tweetMatrix, weighted=T, mode = "undirected")*

> *V(g)$degree <- degree(g)*

> *g <- simplify(g)*

> *V(g)$label <- V(g)$name*

> *V(g)$label.cex <- 1*

> *V(g)$label.color <- rgb(.4, 0, 0, .7)*

> *V(g)$size <- 2*

> *V(g)$frame.color <- NA*

Next, we have a look at the distribution of degree of vertices and the result is shown in the below bar graph as in *Fig. 7.36*. We can see that there are around 20 isolated vertices (with a degree of zero). Note that most of them are caused by the removal of the two keywords, "game" and "play".

> *barplot(table(V(g)$degree))*

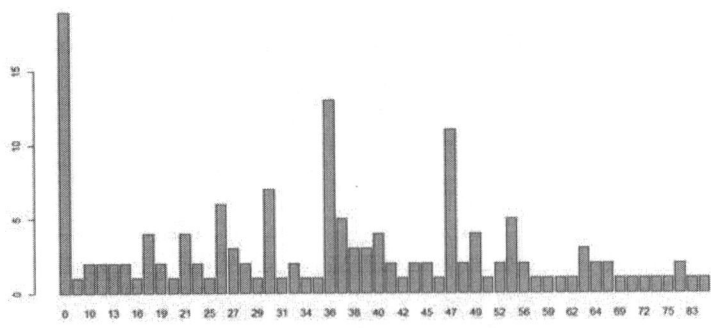

**Figure 7.36**   Distribution of Degree

With the code below, the vertex colours are set based on degree, and labels of isolated vertices are set to tweet IDs and the first 10 characters of every tweet. The labels of other vertices are set to tweet IDs only, so that the graph will not be overcrowded with labels. The colour and width of edges are set based on their weights. The produced graph is shown in *Fig. 7.37.*

```
> idx <- V(g)$degree == 0
> V(g)$label.color[idx] <- rgb(0, 0, .3, .7)
> V(g)$label[idx] <- paste(V(g)$name[idx], substr(dat$text[idx], 1, 10), sep=": ")
> egam <- (log(E(g)$weight)+.2) / max(log(E(g)$weight)+.2)
> E(g)$color <- rgb(.5, .5, 0, egam)
> E(g)$width <- egam
> set.seed(3152)
> layout2 <- layout.fruchterman.reingold(g)
> plot(g, layout=layout2)
```

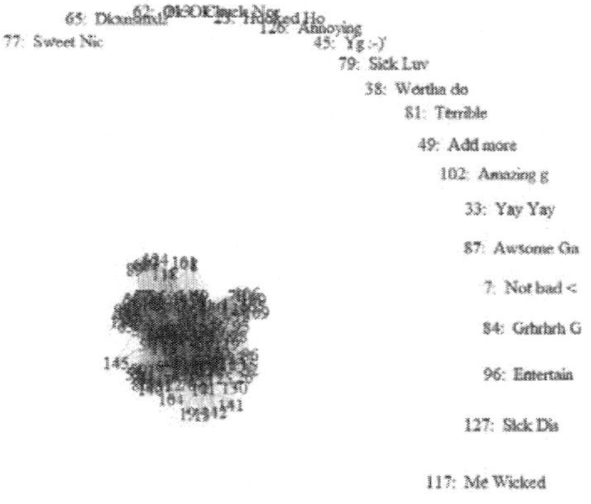

**Figure 7.37**   Network of Tweets - I

The vertices in crescent are isolated from all others, and next they are removed from the graph with the function *delete.vertices()* and re-plot the graph as in *Fig. 7.38.*

```
> g2 <- delete.vertices(g, V(g)[degree(g)==0])
> plot(g2, layout=layout.fruchterman.reingold)
```

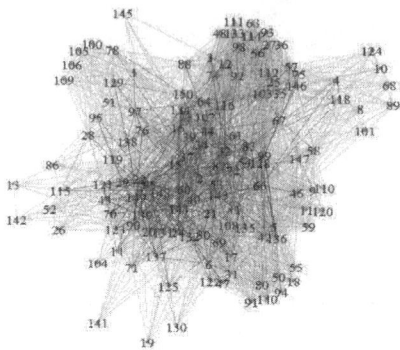

**Figure 7.38** Network of Tweets - II

Similarly, it is also possible to remove the edges with low degrees to simplify the graph using the function *delete.edges()*. After removing edges, some vertices become isolated and they are also removed. The produced graph is as in *Fig. 7.39*.

> g3 <- delete.edges(g, E(g)[E(g)$weight <= 1])

> g3 <- delete.vertices(g3, V(g3)[degree(g3) == 0])

> plot(g3, layout=layout.fruchterman.reingold)

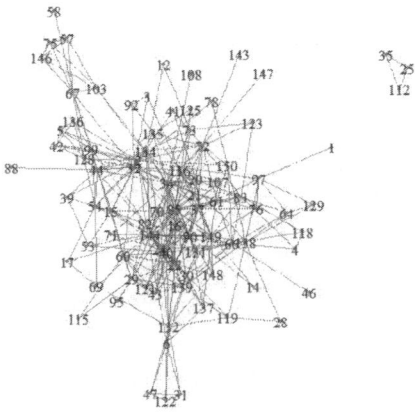

**Figure 7.39** Network of Tweets - III

In the *Fig. 7.39*, there are some groups (or cliques) of tweets. Few of them are listed below. The group of tweets (25, 35, 112) is about the word *"Awesome"*, the group of tweets (31, 47, 122) is about the word *"good"* and the group of tweets (57, 58, 67, 75, 103, 146) is about the word *"addictive"*.

> *dat$text[c(25,35,112)]*

*[1] " Awesome! A lot of fun!!"*

*[2] " Awesome Mysterious Game!! Fun game to play @ night before bed to wind down!!"*

*[3] " Miss Awesome fun"*

> *dat$text[c(31,47,122)]*

*[1] " Error in patching Every time I try to log it it says error in patching but overall good game."*

*[2] " Good For spending time while waiting for an appointment"*

*[3] " Good It is a good game to play while wasting time"*

> *dat$text[c(57,58,67,75,103,146)]*

*[1] " Addictive fun Perfect fun"*

*[2] " Wonderful Is a great game and addictive. Brilliant"*

*[3] " Addictive Great looking, fun game"*

*[4] "ADDICTIVE!!!! This is a fun and easy to play and lose!!"*

*[5] " Very fun Addictive game, similar to a Tomogotchi. You will want to check in on your village and clan. Building, building, building and re-arranging you village. Some battles too. Ver well constructed."*

*[6] " JD Very addictive fun gaming"*

Next we build a *two-mode network*, which is composed of two types of vertices, the tweets and the terms. At first, we generate a graph g directly from the *termDocMatrix*. After that, different colors and sizes are assigned to term vertices and tweet vertices. Then, we set the width and color of the edges. The graph is then plotted with the function *layout.fruchterman.reingold()* as in *Fig. 7.40*.

> *g <- graph.incidence(termDocMatrix, mode=c("all"))*

> *nTerms <- nrow(M)*

```
> nDocs <- ncol(M)

> idx.terms <- 1:nTerms

> idx.docs <- (nTerms+1):(nTerms+nDocs)

> V(g)$degree <- degree(g)

> V(g)$color[idx.terms] <- rgb(0, 1, 0, .5)

> V(g)$size[idx.terms] <- 6

> V(g)$color[idx.docs] <- rgb(1, 0, 0, .4)

> V(g)$size[idx.docs] <- 4

> V(g)$frame.color <- NA

> V(g)$label <- V(g)$name

> V(g)$label.color <- rgb(0, 0, 0, 0.5)

> V(g)$label.cex <- 1.4*V(g)$degree/max(V(g)$degree) + 1

> E(g)$width <- .3

> E(g)$color <- rgb(.5, .5, 0, .3)

> set.seed(958)

> plot(g, layout=layout.fruchterman.reingold)
```

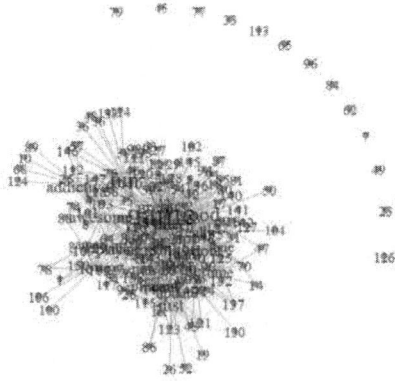

**Figure 7.40** Two-Mode Network of Terms and Tweets - I

The *Fig. 7.40* shows that most tweets are around two centers, *"game"* and *"play"*. Next, let's have a look at which tweets are about *"game"*. In the code below, the function *nei()* returns all vertices which are neighbors of the vertex "game".

```
> V(g)[nei("game")]
+ 89/181 vertices, named:
```
*[1] 2  3  4  5  6  8  9  11  13  15  17  18  20  21  27  28  29*
*[18] 30 31 34 35 37 38 39 40 42 44 51 53 54 55 58 59 60*
*[35] 61 63 64 66 67 71 72 73 76 80 81 82 83 85 87 90 91*
*[52] 92 93 94 95 97 98 99 101 102 103 105 107 108 109 110 111 115*
*[69] 116 117 118 119 120 122 125 127 128 129 131 134 136 138 140 141 143*
*[86] 144 145 148 149*

An alternative way is using the function *neighborhood()* as below.

```
> V(g)[neighborhood(g, order=1, "game")[[1]]]
+ 90/181 vertices, named:
```
*[1] game 2  3  4  5  6  8  9  11  13  15  17  18*
*[14] 20  21  27  28  29  30  31  34  35  37  38  39  40*
*[27] 42  44  51  53  54  55  58  59  60  61  63  64  66*
*[40] 67  71  72  73  76  80  81  82  83  85  87  90  91*
*[53] 92  93  94  95  97  98  99  101 102 103 105 107 108*
*[66] 109 110 111 115 116 117 118 119 120 122 125 127 128*
*[79] 129 131 134 136 138 140 141 143 144 145 148 149*

We can also have a further look at which tweets contain all two terms: *"game"* and *"play"*.

```
> (rdmVertices <- V(g)[nei("game") & nei("play")])
+ 20/181 vertices, named:
```
*[1] 2  6  34 35 37 42 44 59 61 66 73 82 92 107 122 131 134*
*[18] 143 144 149*
```
> dat$text[as.numeric(rdmVertices$label)]
```

[1] " *Great game I love this game. Unlike other games they constantly give you money to play. They are always given you a bone. Keep up the good work.*"

[2] " *Meh Used to be good until World Cup upgrade.\nNow it lags all the time, making it difficult to play.\nMaybe if you spent more time getting the game to actually work and less time trying to squeeze advertising into every nook of game play, we could have a winner.*"
...

Next, we remove "*game*" and "*play*" to show the relationship between tweets with other words. Isolated vertices are also deleted from graph.

```
> idx <- which(V(g)$name %in% c("game", "play"))
> g2 <- delete.vertices(g, V(g)[idx-1])
> g2 <- delete.vertices(g2, V(g2)[degree(g2)==0])
> set.seed(209)
> plot(g2, layout=layout.fruchterman.reingold)
```

From *Fig. 7.41*, we can clearly see groups of tweets and their keywords, such as "*addictive*", "*good*" and "*fun*".

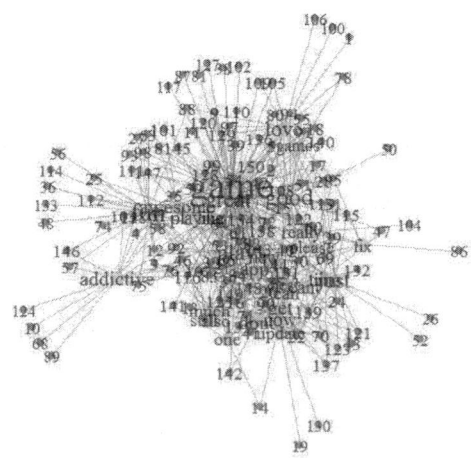

**Figure 7.41 – Two-Mode Network of Terms and Tweets - II**

## ❖ HIGHLIGHTS

- Text mining involves the process of preprocessing the input text, deriving patterns within the preprocessed data, and finally evaluation of the output.

- A word cloud is used to present important words in documents.

- Corpus is a collection of text documents.

- The most accurate and highly used credit scoring measure is the *Probability of Default* called the *PD*.

- The function *importance()* displays the features importance using the *"mean decrease accuracy"* measure.

- Common metrics calculated from the confusion matrix are Precision, Accuracy, True Positive Rate (TP Rate) and False Positive Rate (FP Rate).

- The US crime dataset is used for the EDA of crimes in US

- Pre-processing done in this case study are removing duplicate records, records with missing values, records with incorrect values, formatting Timestamp field, binning time intervals (4 intervals – each 6 hours) and grouping of similar crimes.

- The objective of the EDA on baseball data is to identify trend of base ball players salary over the years, to understand correlation between players salary and their performances, analyze if age, country, height and weight of the players have impact on their performance.

- Social Network Analysis (SNA) is the process of investigating social structures through the use of networks and graph theory.

- The package used for Text Mining is *tm* and the package for Social Network Analysis is *igraph*.

- The function *nei()* returns all vertices which are neighbours of the given vertex.

# GLOSSARY

**ANCOVA**

Analysis of Covariance (ANCOVA) is used to test the main and interaction effects of categorical variables on a continuous dependent variable, controlling for the effects of selected other continuous variables, which co-vary with the dependent.

**ANOVA**

Analysis of variance (ANOVA) is a collection of statistical models and their procedures which are used to observe differences between the means of three or more variables in a population based on the sample presented.

**APRIORI**

The Apriori Algorithm is an influential algorithm for mining frequent itemsets for boolean association rules. Apriori uses a "bottom up" approach, where frequent subsets are extended one item at a time, and groups of candidates are tested against the data.

**Association Rule Mining**

Association rule mining is a procedure which is meant to find frequent patterns, correlations, associations, or causal structures from data sets found in various kinds of databases such as relational databases, transactional databases, and other forms of data repositories.

**Base Environment**

The functions and the variables from the R's base package are stored in the base environment.

**Basic Data Types**

The basic data types in R are Numeric, Integer, Complex, Logical and Character.

| | |
|---|---|
| **Binomial Distribution** | Binomial Distribution is a frequency distribution of the possible number of successful outcomes in a given number of trials in each of which there is the same probability of success. |
| **Chi Square Test** | Chi-square test is a statistical method assessing the goodness of fit between a set of observed values and those expected theoretically. |
| **Classification** | Classification is a data mining function that assigns items in a collection to target categories or classes. The goal of classification is to accurately predict the target class for each case in the data. |
| **Closure** | In R, a function and its environment together is called a closure. |
| **Clustering** | Cluster analysis or clustering is the task of grouping a set of objects in such a way that objects in the same group (called a cluster) are more similar to each other than to those in other groups (clusters). |
| **Confidence** | Confidence is the percentage of cases containing A that also contain B where A and B are two disjoint item sets. $\text{Confidence}(A => B) = P(B|A) = P(A \cup B) / P(A)$ |
| **Confusion matrix** | A confusion matrix is a table that is often used to describe the performance of a classification model (or "classifier") on a set of test data. |
| **Correlation Analysis** | Correlation analysis is a method of statistical evaluation used to study the strength of a relationship between two, numerically measured, continuous variables (e.g. height and weight). |
| **CRAN** | CRAN is a network of ftp and web servers around the world that store identical, up-to-date, versions of code and documentation for R. |

| | |
|---|---|
| **Data Frame** | A data frame is a table or a two-dimensional array-like structure in which each column contains values of one variable and each row contains one set of values from each column. |
| **Decision Tree** | A decision tree is a decision support tool that uses a tree-like graph or model of decisions and their possible consequences, including chance event outcomes, resource costs, and utility. It is one way to display a classification algorithm. |
| **Density Based Clustering** | Clustering based on density (local cluster criterion), such as density-connected points or based on an explicitly constructed density function. |
| **Dimensionality Reduction** | Dimensionality reduction is the process of reducing the number of random variables under consideration, via obtaining a set of principal variables. It can be divided into feature selection and feature extraction. |
| **Document-term matrix** | A document-term matrix or term-document matrix is a mathematical matrix that describes the frequency of terms that occur in a collection of documents. In a document-term matrix, rows correspond to documents in the collection and columns correspond to terms. |
| **Environment** | In R the variables that we create need to be stored in an environment. Environments are another type of variables. We can assign them, manipulate them and pass them as arguments to functions. |
| **Exploratory Data Analysis** | Exploratory Data Analysis (EDA) is an approach to analyzing data sets to summarize their main characteristics, often with visual methods. |

| | |
|---|---|
| **Factors** | Factors in R are stored as a vector of integer values with a corresponding set of character values to use when the factor is displayed. The factor function is used to create a factor. |
| **Feature Selection** | Feature selection, also known as variable selection, attribute selection or variable subset selection, is the process of selecting a subset of relevant features (variables, predictors) for use in model construction. |
| **F-value** | The F critical value is what is referred to as the F statistic. In general, if your calculated F statistic in a test is larger than your table F value, you can reject the null hypothesis. |
| **GitHub** | GitHub is a web-based Git or version control repository and Internet hosting service. It is mostly used for code. |
| **Global Environment** | An environment is just a place to store variables. If we start up R and make an assignment, we are adding an entry in the global environment. |
| **Hierarchical Clustering** | In data mining, hierarchical clustering (also called hierarchical cluster analysis or HCA) is a method of cluster analysis which seeks to build a hierarchy of clusters. |
| **Histogram** | Histogram is a diagram consisting of rectangles whose area is proportional to the frequency of a variable and whose width is equal to the class interval. |
| **Home Directory** | The R home directory is the top-level directory of the R installation being run. The R home directory is often referred to as R_HOME. |

| | |
|---|---|
| **Hypothesis Testing** | Hypothesis Testing is the theory, methods, and practice of testing a hypothesis by comparing it with the null hypothesis. The null hypothesis is only rejected if its probability falls below a predetermined significance level, in which case the hypothesis being tested is said to have that level of significance. |
| **Imputation** | Imputation is the process of replacing missing data with substituted values. |
| **JSON** | JSON (JavaScript Object Notation) is a lightweight data-interchange format. |
| **K-Means Clustering** | K-Means clustering aims to partition n observations into k clusters in which each observation belongs to the cluster with the nearest mean, serving as a prototype of the cluster. This results in a partitioning of the data space into Voronoi cells. |
| **K-Medoids Clustering** | The K-Medoids Clustering algorithm is a clustering algorithm related to the K-Means algorithm. K-Medoids chooses data points as centres and works with a generalization of the Manhattan Norm to define distance between data points. |
| **Least Square Regression** | Linear Least Squares Regression is the most widely used modeling method in which the sum of the squares of the vertical distances of different points from the regression curve is minimized. |
| **Linear Regression** | Linear Regression is an approach for modeling the relationship between a scalar dependent variable y and one or more explanatory variables (or independent variables) denoted X. |
| **Local Outlier Factor** | The local outlier factor (LOF) is an algorithm for finding anomalous data points by measuring the local deviation of a given data point with respect to its neighbours. |

**Logistic Regression**    Logistic regression, is a regression model where the dependent variable (DV) is categorical.

**Lower Tail Test**    In Hypothesis Testing, the alternative hypothesis can take the form - $H1: \mu < \mu 0$ , where a decrease is hypothesized and this is called a lower-tailed test.

**Mean Decrease Accuracy**    Mean Decrease Accuracy is a measure for feature importance in Random Forests. This is assessed for each feature by removing the association between that feature and the target.

**Multiple Regression**    Multiple regression is an extension of simple linear regression. It is used when we want to predict the value of a variable based on the value of two or more other variables. The variable we want to predict is called the dependent variable.

**Multivariate Outlier**    A multivariate outlier is a combination of unusual scores on at least two variables.

**Normal Distribution**    The normal (or Gaussian) distribution is a very common continuous probability distribution. Normal distributions are important in statistics and are often used in the natural and social sciences to represent real-valued random variables whose distributions are not known.

**Null Hypothesis**    A null hypothesis is a type of hypothesis used in statistics that proposes that no statistical significance exists in a set of given observations. The null hypothesis attempts to show that no variation exists between variables or that a single variable is no different than its mean.

| | |
|---|---|
| **One-tailed Test** | In Hypothesis testing, a one-tailed test is appropriate if the estimated value may depart from the reference value in only one direction, for example, whether a machine produces more than one-percent defective products. |
| **Outlier Detection** | In data mining, outlier detection is the identification of items, events or observations which do not conform to an expected pattern or other items in a dataset. |
| **Package** | A Package is a collection of R functions and datasets. Currently, the CRAN package repository features 10756 available packages. |
| **Parent Environment** | All environments are nested and so every environment has a parent environment. The empty environment sits at the top of the hierarchy without any parent. |
| **Poisson Regression** | Poisson regression assumes the response variable Y has a Poisson distribution, and assumes the logarithm of its expected value can be modeled by a linear combination of unknown parameters. |
| **Principal Component Analysis** | Principal component analysis (PCA) is a statistical procedure that uses an orthogonal transformation to convert a set of observations of possibly correlated variables into a set of values of linearly uncorrelated variables called principal components. |
| **Probability of Default** | Probability of default (PD) is a financial term describing the likelihood of a default over a particular time horizon. It provides an estimate of the likelihood that a borrower will be unable to meet its debt obligations. |

**P-value**

When you perform a hypothesis test in statistics, a p-value helps you determine the significance of your results. The p-value is a number between 0 and 1 and if the p-value (typically $\leq 0.05$) indicates strong evidence against the null hypothesis, so we can reject the null hypothesis.

**Quantile**

Quantiles are cutpoints dividing the range of a probability distribution into contiguous intervals with equal probabilities, or dividing the observations in a sample in the same way.

**Regression Analysis**

Regression Analysis is a statistical process for estimating the relationships among variables. It includes many techniques for modeling and analyzing several variables, when the focus is on the relationship between a dependent variable and one or more independent variables (or 'predictors').

**R-Forge**

R-Forge strives to provide a collaborative environment for R package developers.

**Scatter Plot**

A scatter plot is a type of plot using Cartesian coordinates to display values for typically two variables for a set of data. The data is displayed as a collection of points, each having the value of one variable determining the position on the horizontal axis and the value of the other variable determining the position on the vertical axis.

**Significance Level**

The significance level, also denoted as alpha or $\alpha$, is the probability of rejecting the null hypothesis when it is true. For example, a significance level of 0.05 indicates a 5% risk of concluding that a difference exists when there is no actual difference.

| | |
|---|---|
| Silhouette | Silhouette analysis can be used to study the separation distance between the resulting clusters. The silhouette plot displays a measure of how close each point in one cluster is to points in the neighboring clusters and thus provides a way to assess parameters like number of clusters visually. |
| SMOTE | The SMOTE function handles unbalanced classification problems and it can generate a new "SMOTEd" data set that addresses the class unbalance problem. |
| Social Network Analysis | Social Network Analysis (SNA) is the process of investigating social structures through the use of networks and graph theory. |
| Stemming | Stemming is the process of reducing inflected (or sometimes derived) words to their word stem, base or root form, generally a written word form. |
| Stop-word | A Stop Word is a commonly used word (such as "the") that a search engine has been programmed to ignore, both when indexing entries for searching and when retrieving them as the result of a search query. |
| Support | Support is the percentage of cases in the data that contains both A and B, where A and B are two disjoint item sets. Support$(A => B) = P(A \cup B)$ |
| Term-document Matrix | A document-term matrix or term-document matrix is a mathematical matrix that describes the frequency of terms that occur in a collection of documents. In a document-term matrix, rows correspond to documents in the collection and columns correspond to terms. |
| Term-term Adjacency Matrix | A Term-term Adjacency Matrix is a square matrix used to represent a finite graph in which the elements of the matrix indicate whether pairs of terms (vertices) are adjacent or not in the graph. |

| | |
|---|---|
| **Text Mining** | Text mining, is the process of deriving high-quality information from text, which is derived through the devising of patterns and trends such as statistical pattern learning or data mining. |
| **Transpose** | The transpose of a matrix is an operator which flips a matrix over its diagonal, that is it switches the row and column indices of the matrix by producing another matrix. |
| **Two-tailed test** | A two-tailed test is a statistical test in which the critical area of a distribution is two-sided and tests whether a sample is greater than or less than a certain range of values. If the sample being tested falls into either of the critical areas, the alternative hypothesis is accepted instead of the null hypothesis. |
| **Type I error** | In hypothesis testing, a type I error is the incorrect rejection of a true null hypothesis (a "false positive"). |
| **Type II error** | In hypothesis testing, a type II error is incorrectly retaining a false null hypothesis (a "false negative"). |
| **Univariate Outlier** | A univariate outlier is a data point that consists of an extreme value on one variable. Outliers can influence the outcome of statistical analyses. |
| **Upper Tail Test** | In Hypothesis Testing, the alternative hypothesis can take the form - H1: $\mu > \mu 0$ , where $\mu 0$ is the comparator or null value and an increase is hypothesized - this type of test is called an upper-tailed test. |
| **Wordcloud** | A word cloud is a visual representation of text data, typically used to depict keyword on websites, or to visualize free form text in which the importance of each word is shown with font size or color. |

# PACKAGES USED

| Package | Description | Page |
|---|---|---|
| **party** | A Package for Recursive Partytioning | 113, 122, 123, 151 |
| **plotrix** | Package that contains Various Plotting Functions | 73, 86, 87 |
| **plyr** | Package with Tools for Splitting, Applying and Combining Data | 58, 60, 113, 147, 151 |
| **randomForest** | Package for Random Forests for Classification and Regression | 130, 149, 151, 160, 165 |
| **Rcpp** | Package for R and C++ Integration | 171, 181 |
| **reshape2** | Package that Flexibly Reshapes Data | 63 |
| **rpart** | Package for Recursive Partitioning and Regression Trees | 127, 130, 151, 161, 167 |
| **RSQLite** | Package that provides SQLite interface for R | 55 |

| Package | Description | Page |
|---|---|---|
| **sqldf** | Package that Performs SQL Selects on R Data Frames | 52, 60, 61, 70 |
| **stats** | The R Package for Statistics | 112, 113, 114, 136, 151 |
| **stringr** | Simple, Consistent Wrappers for Common String Operations | 56 |
| **TH.data** | Package for TH's Data Archive | 125 |

# FUNCTIONS USED

| Function | Package | Description | Page |
|----------|---------|-------------|------|
| **.libPaths()** | base | .libPaths gets/sets the library trees within which packages are looked for. | 7 |
| **abline()** | graphics | This function adds one or more straight lines through the current plot. | 100, 102, 126 |
| **acos()** | base | This function computes the arc-cosine. | 6 |
| **aes()** | ggplot2 | Aesthetic mappings describe how variables in the data are mapped to visual properties (aesthetics) of geoms. | 86, 155, 174, 182, 183 |
| **aggregate()** | stats | Splits the data into subsets, computes summary statistics for each, and returns the result in a convenient form. | 63, 70, 171,172, 174, |
| **all.equal()** | base | all.equal(x, y) is a utility to compare R objects x and y testing 'near equality'. | 6 |
| **anova()** | stats | Compute analysis of variance (or deviance) tables for one or more fitted model objects. | 107, 108, 109, 112 |

| Function | Package | Description | Page |
|----------|---------|-------------|------|
| **aov()** | stats | Fit an analysis of variance model by a call to lm for each stratum. | 107,108, 109, 112, |
| **apply()** | base | Returns a vector or array or list of values obtained by applying a function to margins of an array or matrix. | 63, 70, 145 |
| **apriori()** | arules | This function Mines frequent itemsets, association rules or association hyperedges using the Apriori algorithm. | 131, 134,151, |
| **args()** | base | Displays the argument names and corresponding default values of a function or primitive. | 10.11 |

| Function | Package | Description | Page |
|----------|---------|-------------|------|
| **arrange()** | plyr | This function completes the subsetting, transforming and ordering triad with a function that works in a similar way to subset and transform but for reordering a data frame by its columns. | 60,149 |
| **array()** | base | Creates or tests for arrays. | 23, 64 |
| **as.character()** | base | Create or test for objects of type "character". | 19,30,172, 173 |
| **as.data.frame()** | base | Functions to check if an object is a data frame, or coerce it if possible. | 130 |

| Function | Package | Description | Page |
|---|---|---|---|
| **as.data.table()** | data. table | Functions to check if an object is data.table, or coerce it if possible. | 180, 182 |
| **as.Date()** | base | Functions to convert between character representations and objects of class "Date" representing calendar dates. | 43, 45, 49, 58 |
| **as.environment ()** | base | A generic function coercing an R object to an environment. | 9 |
| **as.integer()** | basc | Creates or tests for objects of type "integer". | 18 |
| **as.list()** | base | Functions to construct, coerce and check for both kinds of R lists. | 9, 29, 30 |
| **as.numeric()** | base | Creates or coerces objects of type "numeric". is.numeric is a more general test of an object being interpretable as numbers. | 30, 165, 182, 183 |
| **as.POSIXlt()** | base | Functions to manipulate objects of classes "POSIXlt" and "POSIXct" representing calendar dates and times. | 42, 43, 44, 170, 172 |
| **as.yaml()** | yaml | Convert an R object into a YAML string | 53 |
| **asin()** | base | This function computes the arc-sine. | 6 |

| Function | Package | Description | Page |
|---|---|---|---|
| **assign()** | base | Assign a value to a name in an environment. | 4, 9,16 |
| **atan()** | base | This function computes the arc-tan. | 6 |

| Function | Package | Description | Page |
|---|---|---|---|
| **available. packages()** | utils | Returns a matrix of details corresponding to packages currently available at one or more repositories. | 8 |
| **BarChart()** | lattice | Plots a bar chart, such as for counts, with default colors for one or two variables. | 85,87 |
| **barplot()** | graphics | Creates a bar plot with vertical or horizontal bars. | 85, 87, 155 |
| **basename()** | base | Removes all of the path up to and including the last path separator. | 42 |
| **biplot()** | stats | Plot a biplot on the current graphics device. | 140, 151 |
| **body()** | base | Get or set the body of a function. | 10, 11 |

| Function | Package | Description | Page |
|---|---|---|---|
| **boxplot()** | graphics | Produce box-and-whisker plot(s) of the given (grouped) values. | 87, 137, 162, 163 |
| **boxplot. stats()** | grDevices | This function is typically called by another function to gather the statistics necessary for producing box plots, but may be invoked separately. | 136, 137, 151 |
| **brewer.pal()** | RColorBrewer | Creates nice looking color palettes especially for thematic maps. | 156 |
| **bwplot()** | lattice | Produces box-and-whisker plots | 82,87 |
| **by()** | base | Function by is an object-oriented wrapper for tapply applied to data frames. | 63,67, 68, 70, |
| **c()** | base | This is a generic function which combines its arguments. | 5, 16, 22, 30, 39 |
| **cat()** | base | Outputs the objects, concatenating the representations. cat performs much less conversion thanprint. | 39,40, 159 |

| Function | Package | Description | Page |
|----------|---------|-------------|------|
| **cbind()** | base | Take a sequence of vector, matrix or data-frame arguments and combine by *columns* or *rows*, respectively. | 25, 49, 61, 130 |
| **ceiling_date()** | lubridate | Takes a date-time object and rounds it up to the nearest boundary of the specified time unit. | 48 |

| Function | Package | Description | Page |
|----------|---------|-------------|------|
| **chisq.test()** | stats | Performs chi-squared contingency table tests and goodness-of-fit tests. | 109, 110, 112 |
| **clara()** | cluster | Computes a "clara" object, a list representing a clustering of the data into k clusters. | 116, 151 |
| **class()** | base | A simple generic function mechanism which can be used for an object-oriented style of programming. | 18,43 |
| **cmdscale()** | stats | Classical multidimensional scaling (MDS) of a data matrix. Also known as principal coordinates analysis. | 164 |
| **coef()** | stats | A generic function which extracts model coefficients from objects returned by modeling functions. | 101 |

| Function | Package | Description | Page |
|---|---|---|---|
| **cohesive. blocks()** | igraph | Calculates cohesive blocks for objects of class igraph. | 191 |
| **colMeans()** | base | Form row and column sums and means for numeric arrays. | 35, 36, 147 |
| **colnames()** | base | Retrieve or set the row or column names of a matrix-like object. | 24, 31, 59, 73, 88 |
| **colSums()** | base | Form row and column sums and means for numeric arrays. | 35, 36, 133, |
| **complete. cases()** | stats | Return a logical vector indicating which cases are complete, i.e., have no missing values. | 59 |
| **confint()** | stats | Computes confidence intervals for one or more parameters in a fitted model. | 106 |
| **content_ transformer()** | tm | Create content transformers, i.e., functions which modify the content of an R object. | 153 |
| **coord_flip()** | ggplot2 | Flip cartesian coordinates so that horizontal becomes vertical, and vertical, horizontal. | 86, 155 |
| **cor()** | stats | Computes the correlation of x and y if these are vectors. | 91, 96, 112, 165 |

| Function | Package | Description | Page |
|---|---|---|---|
| **cor.test()** | stats | Test for association between paired samples, using one of Pearson's product moment correlation coefficient, Kendall's *tau* or Spearman's *rho*. | 96, 97, 112 |
| **Corpus()** | tm | Representing and computing on corpora. | 153 |
| **corrplot()** | corrplot | A graphical display of a correlation matrix, confidence interval. | 97, 112, 148, |
| **cov()** | stats | Compute the covariance of x and y if these are vectors. | 91 |
| **ctree()** | party | Recursive partitioning for continuous, censored, ordered, nominal and multivariate response variables in a conditional inference framework. | 122, 123, 124, 130, 151 |
| **cummax()** | base | Returns a vector whose elements are the cumulative maxima of the elements of the argument. | 91 |
| **cummin()** | base | Returns a vector whose elements are the cumulative minima of the elements of the argument. | 91 |
| **cumprod()** | base | Returns a vector whose elements are the cumulativeproducts of the elements of the argument. | 91 |
| **cumsum()** | base | Returns a vector whose elements are the cumulative sums of the elements of the argument. | 91 |

| Function | Package | Description | Page |
|---|---|---|---|
| **cut()** | base | Divides the range of x into intervals and codes the values in x according to which interval they fall. | 38, 171, 172 |
| **cutree()** | stats | Cuts a tree, as resulting from hclust, into several groups either by specifying the desired number(s) of groups or the cut height(s). | 118, 119, 157 |
| **daisy()** | cluster | Compute all the pairwise dissimilarities (distances) between observations in the data set. | 114, 160, 163, 164, |

| Function | Package | Description | Page |
|---|---|---|---|
| **data()** | utils | Loads specified data sets, or list the available data sets. | 50, 74 |
| **data.frame()** | base | This function creates data frames, tightly coupled collections of variables which share many of the properties of matrices and of lists, used as the fundamental data structure by most of R's modeling software. | 11, 32, 58, 60, 61, 100, 130, 137, 149, 155, 167, 168 |
| **days()** | chron | Returns Various Periods from a Chron or Dates Object | 47, 49 |
| **dbConnect()** | dbConnect | Creates a connection to a DBMS | 55 |

| Function | Package | Description | Page |
|----------|---------|-------------|------|
| **dbDisconnect()** | DBI | Function to disconnect from a database | 55 |
| **dbDriver()** | DBI | Function to create a new driver object given the name of a database | 55 |
| **dbGetQuery()** | DBI | Function to Send query, retrieve results and then clear result set. Returns the result of a query as a data frame. | 55 |
| **dbinom()** | stats | Density, distribution function, quantile function and random generation for the binomial distribution with parameters size and prob. | 95, 112 |
| **dbListTables()** | DBI | This function returns a character vector that enumerates all tables and views in the database. | 55 |
| **dbReadTable()** | DBI | This Function returns a data frame that contains the complete data from the remote table. | 55 |
| **dbscan()** | fpc | Generates a density based clustering of arbitrary shape. | 114, 119, 120, 141, 151 |
| **dbUnloadDriver()** | DBI | Function to remove a driver object given the name of a database | 55 |

| Function | Package | Description | Page |
|----------|---------|-------------|------|
| **dcast()** | reshape2 | Cast functions Cast a molten data frame into an array or data frame. | |
| **ddays()** | lubridate | Quickly creates Duration days for easy date-time manipulation. | 46, 47 |

| Function | Package | Description | Page |
|----------|---------|-------------|------|
| **degree()** | igraph | The degree of a vertex is its most basic structural property, the number of its adjacent edges. | 190, 194, 195, 196, 197, 198 |
| **delete.edges()** | igraph | Deletes edges from a graph. | 196 |
| **delete. vertices()** | igraph | Deletes vertices from a graph. | 195, 196, 198 |
| **density()** | stats | The function density computes kernel density estimates. Its default method does so with the given kernel and bandwidth for univariate observations. | 139 |
| **deparse()** | base | Turn unevaluated expressions into character strings. | 10, 11 |
| **dhours()** | lubridate | Quickly creates Duration hours for easy date-time manipulation. | 46, 49 |
| **difftime()** | base | Time intervals creation, printing, and some arithmetic. | 45 |

| Function | Package | Description | Page |
|---|---|---|---|
| **dim**() | base | Retrieve or set the dimension of an object. | 24, 27, 118, 144, 147, 148 |
| **dimnames**() | base | Retrieve or set the dimnames of an object. | 24, 31, 154, |
| **dirname**() | base | basename removes all of the path up to and including the last path separator. | 42 |
| **dist**() | stats | This function computes and returns the distance matrix computed by using the specified distance measure to compute the distances between the rows of a data matrix. | 118, 157, |
| **dminutes**() | lubridate | Quickly creates Duration minutes for easy date-time manipulation. | 46, 49 |
| **dmy**() | lubridate | Parse dates according to the order in that day, month, and year elements appear in the input vector.Transforms dates stored in character and numeric vectors to Date or POSIXct objects. | 46 |

| Function | Package | Description | Page |
|---|---|---|---|
| **dnorm()** | stats | Density, distribution function, quantile function and random generation for the normal distribution with mean equal to mean and standard deviation equal to sd. | 92, 93 |
| **do.call()** | base | do.call constructs and executes a function call from a name or a function and a list of arguments to be passed to it. | 11 |
| **download.file()** | utils | This function can be used to download a file from the Internet. | 54 |
| **droplevels()** | base | The function droplevels is used to drop unused levels from a factor or, more commonly, from factors in a data frame. | 37 |
| **dseconds()** | lubridate | Quickly creates Duration seconds for easy date-time manipulation. | 46, 49 |
| **dweeks()** | lubridate | Quickly creates Duration weeks for easy date-time manipulation. | 46, 47, 49 |
| **dyears()** | lubridate | Quickly creates Duration years for easy date-time manipulation. | 46, 47, 49 |
| **dym()** | lubridate | Parse dates according to the order in that day, year, and month elements appear in the input vector. | 46 |

| Function | Package | Description | Page |
|---|---|---|---|
| eclat() | arules | This funciton Mines frequent itemsets with the Eclat algorithm. | 131, 151 |
| exists() | base | Look for an R object of the given name and possibly return it | 9, 10 |
| facet_wrap() | ggplot2 | This function wraps a 1d sequence of panels into 2d. This is generally a better use of screen space than facet_grid because most displays are roughly rectangular. | 77, 87 |
| factor() | base | The function factor is used to encode a vector as a factor (the terms 'category' and 'enumerated type' are also used for factors). | 36, 66, 139, 149, 151, 163, 164, 171 |

| Function | Package | Description | Page |
|---|---|---|---|
| file.path() | base | Construct the path to a file from components in a platform-independent way. | 42 |
| findAssocs() | tm | Find associations in a document-term or term-document matrix. | 156 |
| findCorrelation() | caret | This function searches through a correlation matrix and returns a vector of integers corresponding to columns to remove to reduce pair-wise correlations. | 148 |

| Function | Package | Description | Page |
|----------|---------|-------------|------|
| **findFreqTerms()** | tm | Find frequent terms in a document-term or term-document matrix. | 155 |
| **fixed()** | stringr | Compare literal bytes in the string. This is very fast, but not usually what you want for non-ASCII character sets. | 56, 57 |
| **floor_date()** | lubridate | This function takes a date-time object and rounds it down to the nearest boundary of the specified time unit. | 48 |
| **force_tz()** | lubridate | This function returns a the date-time that has the same clock time as x in the new time zone. | 48 |
| **formalArgs()** | methods | These are utilities, currently in the methods package, that either provide some functionality needed by the package. | 10,11 |
| **formals()** | base | Get or set the formal arguments of a function. | 10 |
| **format()** | base | Format an R object for pretty printing. | 40, 46, 171 |
| **formatC()** | base | Formatting numbers individually and flexibly, formatC() using C style format specifications. | 40 |
| **fromJSON()** | jsonlite | This function is used to convert between JSON data and R objects. | 53 |

| Function | Package | Description | Page |
|---|---|---|---|
| **geom_bar()** | ggplot2 | This function makes the height of the bar proportional to the number of cases in each group. | 86, 87, 155, |

| Function | Package | Description | Page |
|---|---|---|---|
| **grep()** | base | Search for matches to argument patternwithin each element of a character vector: they differ in the format of and amount of detail in the results. | 56, 70, 154 |
| **grepl()** | base | Search for matches to argument patternwithin each element of a character vector: they differ in the format of and amount of detail in the results. | 56 |
| **gsub()** | base | Search for matches to argument patternwithin each element of a character vector: they differ in the format of and amount of detail in the results. | 56, 70 |
| **hclust()** | stats | Hierarchical cluster analysis on a set of dissimilarities and methods for analyzing it. | 114, 118, 119, 151, 157, |
| **head()** | utils | Returns the first or last parts of a vector, matrix, table, data frame or function. Since head() andtail() are generic functions, they may also have been extended to other classes. | 32, 49, 50, 52, 62, 137, |

| Function | Package | Description | Page |
|---|---|---|---|
| **help()** | utils | help is the primary interface to the help systems. | 4, 16 |
| **help. search()** | utils | Allows for searching the help system for documentation matching a given character string in the (file) name, alias, title, concept or keyword entries (or any combination thereof), using either fuzzy matching or regular expression matching. | 4 |
| **hist()** | graphics | The generic function hist computes a histogram of the given data values. If plot = TRUE, the resulting object of class "histogram" is plotted by plot.histogram, before it is returned. | 80, 81, 87, 95 |
| **histogram()** | lattice | Draw Histograms, possibly conditioned on other variables. | 183 |
| **hours()** | chron | Return Hours from a Times Object. | 47, 49 |

| Function | Package | Description | Page |
|---|---|---|---|
| **htmlParse()** | XML | This function Parses a HTML file or HTML content, and generates an R structure representing the HTML tree. | 53 |
| **htmlTreeParse()** | XML | Parses an HTML file or string containing HTML content, and generates an R structure representing the HTML tree. | 53 |

| Function | Package | Description | Page |
|---|---|---|---|
| **identical()** | base | The safe and reliable way to test two objects for being *exactly* equal. It returns TRUE in this case, FALSE in every other case. | 5, 60 |
| **ifelse()** | base | Returns a value with the same shape as test which is filled with elements selected from either yes or no depending on whether the element of test is TRUE or FALSE. | 13, 16, 164, 171, 172, 173, |
| **importance()** | random Forest | This is the extractor function for variable importance measures as produced by randomForest. | 129, 151, 165, 166 |
| **inspect()** | arules | Provides the generic function inspect to display associations plus additional information formatted for online inspection. | 131, 132, 133, 134, 154 |
| **install. packages()** | utils | Download and install packages from CRAN-like repositories or from local files. | 8, 16, 46, 51, 73, 171, 181 |
| **installed. packages()** | utils | Find (or retrieve) details of all packages installed in the specified libraries. | 7 |
| **interaction()** | base | Computes a factor which represents the interaction of the given factors. The result of interaction is always unordered. | 39, 49 |

| Function | Package | Description | Page |
|----------|---------|-------------|------|
| intersect() | base | Performs the set operation intersect on the two member vectors | 138 |
| is.atomic() | base | is.atomic returns TRUE if x is of an atomic type (or NULL) and FALSE otherwise. | 27 |
| is.integer() | base | Creates or tests for objects of type "integer". | 18 |

| Function | Package | Description | Page |
|----------|---------|-------------|------|
| is.na() | base | NA is a logical constant of length 1 which contains a missing value indicator. NA can be coerced to any other vector type except raw. | 37, 147, 162, 170, 172 |
| is.recursive() | base | is.atomic returns TRUE if x is of an atomic type (or NULL) and FALSE otherwise. | 27 |
| isTRUE() | base | These operators act on raw, logical and number-like vectors. | 6 |
| kmeans() | stats | Perform k-means clustering on a data matrix. | 114, 115, 143, 151, 158 |
| knnImputation() | DMwR | Function that fills in all NA values using the k Nearest Neighbours of each case with NA values. | 160 |

| Function | Package | Description | Page |
|---|---|---|---|
| **lapply()** | base | lapply returns a list of the same length as X, each element of which is the result of applying FUNto the corresponding element of X. | 63, 64, 65, 153, |
| **largest.cliques()** | igraph | This function finds all, the largest cliques in an undirected graph. The size of the largest clique can also be calculated. | 193 |
| **layout()** | base | layout divides the device up into as many rows and columns as there are in matrix mat, with the column-widths and the row-heights specified in the respective arguments. | 74, 87, 117, |
| **layout. fruchterman. reingold()** | igraph | This is a deprecated layout function. | 190, 194, 197 |
| **legend()** | graphics | This function can be used to add legends to plots. Note that a call to the function locator(1) can be used in place of the x and y arguments. | 72, 79, 84, 85 |
| **length()** | base | Get or set the length of vectors (including lists) and factors, and of any other R object for which a method has been defined. | 21, 24, 31, 56, 65, 72, 107, 173 |

| Function | Package | Description | Page |
|----------|---------|-------------|------|
| **match()** | base | match returns a vector of the positions of (first) matches of its first argument in its second. | 90, 112 |
| **matrix()** | base | matrix creates a matrix from the given set of values. | 23, 25, 63, 74, 117, 121 |
| **max()** | base | Returns the (parallel) maxima and minima of the input values. | 88, 91, 112, 156, 182 |
| **maximal. cliques()** | igraph | This function finds all the maximal cliques in an undirected graph. | 193 |
| **mdy()** | lubridate | Parse dates according to the order in that month, day, and year elements appear in the input vector. | 46 |
| **mean()** | base | Generic function for the (trimmed) arithmetic mean. | 5, 16, 45, 46, 88, 89, 180, 182 |
| **median()** | stats | Compute the sample median. | 5, 16, 88, 89 |
| **melt()** | reshape2 | Convert an object into a molten data frame. | 62, 63, 70 |
| **merge()** | base | Merge two data frames by common columns or row names, or do other versions of database *join*operations. | 35, 49, 62, 180, 182 |
| **message()** | base | Generate a diagnostic message from its arguments. | 12, 13, 14, 15 |

| Function | Package | Description | Page |
|---|---|---|---|
| **min()** | base | Returns the (parallel) maxima and minima of the input values. | 88,91, 112, 162, 180, 182 |
| **minutes()** | chron | Return Minutes from a Times Object. | 47 |
| **months()** | base | This function Extracts the month. | 171, 172 |
| **mutate()** | plyr | Mutate a data frame by adding new or replacing existing columns. | 58 |
| **myd()** | lubridate | Parse dates according to the order in that month, year, and day elements appear in the input vector. | 46 |

| Function | Package | Description | Page |
|---|---|---|---|
| **na.fail()** | stats | Returns the object if it does not contain any missing values, and signals an error otherwise. | 59 |
| **na.omit()** | stats | Returns the object with incomplete cases removed. | 59, 70 |
| **names()** | base | Functions to get or set the names of an object. | 2, 26, 31, 131, 155, 174 |
| **ncol()** | base | Returns the number of columns present in x. | 24, 27, 31 |

| Function | Package | Description | Page |
|---|---|---|---|
| **nearZeroVar()** | caret | This function diagnoses predictors that have one unique value or predictors that have both of the following characteristics: they have very few unique values relative to the number of samples and the ratio of the frequency of the most common value to the frequency of the second most common value is large. | 147 |
| **nei()** | igraph | This function finds the vertices not farther than a given limit from another fixed vertex, these are called the neighborhood of the vertex. | 198 |
| **neighborhood()** | igraph | This function finds the vertices not farther than a given limit from another fixed vertex, these are called the neighborhood of the vertex. | 198 |
| **new.env()** | base | Get, set, test for and create environments. | 8, 9, 16 |
| **nlevels()** | base | Return the number of levels which its argument has. | 36 |
| **nls()** | stats | Determine the nonlinear (weighted) least-squares estimates of the parameters of a nonlinear model. | 105, 106 |
| **noquote()** | base | Print character strings without quotes. | 40 |

| Function | Package | Description | Page |
|---|---|---|---|
| **nrow()** | base | Returns the number of rows present in x. | 24, 59, 128, 140, 162 |
| **odbcClose()** | RODBC | Close connections to ODBC databases. | 55 |
| **odbcConnect()** | RODBC | Open connections to ODBC databases. | 55 |

| Function | Package | Description | Page |
|---|---|---|---|
| **order()** | base | Returns a permutation which rearranges its first argument into ascending or descending order, breaking ties by further arguments. sort.list is the same, using only one argument. | 61,70, 143, 180, |
| **outliers. ranking()** | DMwR | This function uses hierarchical clustering to obtain a ranking of outliers for a set of cases. | 160, 163 |
| **pairs()** | graphics | A matrix of scatterplots is produced. | 75, 87, 140 |
| **pam()** | cluster | Partitioning (clustering) of the data into k clusters "around medoids", a more robust version of K-means. | 114, 116, 117, 118, 151, |
| **pamk()** | fpc | Partitioning around medoids with estimation of number of clusters | 114, 116, 117 |

| Function | Package | Description | Page |
|---|---|---|---|
| **par()** | graphics | Used to set or query graphical parameters. Parameters can be set by specifying them as arguments to par. | 84, 85 |
| **parent.frame()** | base | These functions provide access to environments ('frames' in S terminology) associated with functions further up the calling stack. | 10 |
| **paste()** | base | Concatenate vectors after converting to character. | 19, 39, 49, 166, 182, 183 |
| **paste0()** | base | Concatenate vectors after converting to character. | 39 |
| **path. expand("~")** | base | Expand a path name, for example by replacing a leading tilde by the user's home directory (if defined on that platform). | 7 |
| **pbinom()** | stats | Density, distribution function, quantile function and random generation for the binomial distribution with parameters size and prob. | 95, 96, 112 |
| **pie()** | graphics | Draw a pie chart. | 72, 87 |
| **pie3D()** | plotrix | Displays a 3D pie chart with optional labels. | 73, 87 |

| Function | Package | Description | Page |
|----------|---------|-------------|------|
| **plot()** | graphics | Generic function for plotting of R objects. For more details about the graphical parameter arguments, see par. | 73, 100, 126, 142, 157, 164, 166, |
| **plotcluster()** | fpc | Plots to distinguish given classes by ten available projection methods. One-dimensional data is plotted against the cluster number. | 120 |
| **plotcorr()** | ellipse | This function plots a correlation matrix using ellipse-shaped glyphs for each entry. The ellipse represents a level curve of the density of a bivariate normal with the matching correlation. | 160, 165 |
| **pmax()** | base | Returns the (parallel) maxima and minima of the input values. | 91 |
| **pmin()** | base | Returns the (parallel) maxima and minima of the input values. | 91 |
| **pnorm()** | stats | Density, distribution function, quantile function and random generation for the normal distribution with mean equal to mean and standard deviation equal to sd. | 92, 93, 112 |
| **prcomp()** | stats | Performs a principal components analysis on the given data matrix and returns the results as an object of class prcomp. | 140, 145, 151 |

| Function | Package | Description | Page |
|----------|---------|-------------|------|
| **predict**() | stats | predict is a generic function for predictions from the results of various model fitting functions. The function invokes particular methods which depend on the class of the first argument. | 99, 100, 122, 123, 124, 125, 126, 128, 129, 168 |
| **print**() | base | print prints its argument and returns it *invisibly* (via invisible(x)). It is a generic function which means that new printing methods can be easily added for new classes. | 4, 61, 124, 126, 127, 139, 143, 149 |
| **printcp**() | rpart | Displays the cp table for fitted rpart object. | 167 |

| Function | Package | Description | Page |
|----------|---------|-------------|------|
| **rbinom**() | stats | Density, distribution function, quantile function and random generation for the binomial distribution with parameters size and prob. | 95, 96, 112 |
| **read.csv**() | utils | Reads a file in table format and creates a data frame from it, with cases corresponding to lines and variables to fields in the file. | 52, 54, 59, 162, 170, 171, 172, 180, 182 |

| Function | Package | Description | Page |
|---|---|---|---|
| **read.csv2()** | utils | Reads a file in table format and creates a data frame from it, with cases corresponding to lines and variables to fields in the file. | 52 |
| **read.delim()** | utils | Reads a file in table format and creates a data frame from it, with cases corresponding to lines and variables to fields in the file. | 52 |
| **read.delim2()** | utils | Reads a file in table format and creates a data frame from it, with cases corresponding to lines and variables to fields in the file. | 52 |
| **read.spss()** | foreign | This function reads a file stored by the SPSS save or export commands. | 54, 70 |
| **read.ssd()** | foreign | Generates a SAS program to convert the ssd contents to SAS transport format and then uses read.xport to obtain a data frame. | 54, 70 |
| **read.table()** | utils | Reads a file in table format and creates a data frame from it, with cases corresponding to lines and variables to fields in the file. | 51. 52, 54 |
| **read.xlsx()** | xlsx | Read the contents of a worksheet into an R data.frame. | 53, 70 |
| **read.xlsx2()** | xlsx | Read the contents of a worksheet into an R data.frame. | 53, 54, 70 |
| **readLines()** | base | Read some or all text lines from a connection. | 10, 52, 70 |

| Function | Package | Description | Page |
|---|---|---|---|
| **readMat()** | R.matlab | Reads matlab files | 54, 70 |
| **rect.hclust()** | stats | Draws rectangles around the branches of a dendrogram highlighting the corresponding clusters. First the dendrogram is cut at a certain level, then a rectangle is drawn around selected branches. | 118, 119, 157 |
| **rclevel()** | stats | The levels of a factor are re-ordered so that the level specified by ref is first and the others are moved down. This is useful for contr. treatment contrasts which take the first level as the reference. | 37 |
| **remove.packages()** | utils | Removes installed packages/bundles and updates index information as necessary. | 8 |
| **remove Sparse Terms()** | tm | Remove sparse terms from a document-term or term-document matrix. | 82 |
| **reorder()** | stats | reorder is a generic function. The "default" method treats its first argument as a categorical variable, and reorders its levels based on the values of a second variable, usually numeric. | 82, 83 |

| Function | Package | Description | Page |
|----------|---------|-------------|------|
| **rep()** | base | rep replicates the values in x. It is a generic function, and the (internal) default method is described here. | 22, 23, 130, 140, 146, |
| **replicate()** | base | lapply returns a list of the same length as X, each element of which is the result of applying FUN to the corresponding element of X. | 12 |
| **require()** | base | library and require load and attach add-on packages. | 7, 163 |
| **reshape()** | stats | This function reshapes a data frame between 'wide' format with repeated measurements in separate columns of the same record and 'long' format with the repeated measurements in separate records. | |

| Function | Package | Description | Page |
|----------|---------|-------------|------|
| **resid()** | stats | residuals is a generic function which extracts model residuals from objects returned by modeling functions. | 106 |

| Function | Package | Description | Page |
|---|---|---|---|
| **rfcv**() | randomForest | This function shows the cross-validated prediction performance of models with sequentially reduced number of predictors (ranked by variable importance) via a nested cross-validation procedure. | 160, 166 |
| **rgb**() | grDevices | This function creates colors corresponding to the given intensities (between 0 and max) of the red, green and blue primaries. | 191, 192, 193, 194, 197 |
| **rnorm**() | stats | Density, distribution function, quantile function and random generation for the normal distribution with mean equal to mean and standard deviation equal to sd. | 12, 92, 94, 95, 112, 136, 137 |
| **round**() | base | This function rounds the values in its first argument to the specified number of decimal places (default 0). | 131, 132, 164, 182, 183 |
| **rowMeans**() | base | Form row and column sums and means for numeric arrays. | 35, 36 |
| **rownames**() | base | Retrieve or set the row or column names of a matrix-like object. | 24, 31, 145, 149 |
| **rowSums**() | base | Form row and column sums and means for numeric arrays. | 35, 36, 143, 155 |

| Function | Package | Description | Page |
|----------|---------|-------------|------|
| **rpart()** | rpart | Recursive Partitioning and Regression Trees. Fit a rpart model. | 122, 125, 127, 130, 151, 161, 167 |
| **runif()** | stats | This function provide information about the uniform distribution on the interval from min to max. The function runif generates random deviates. | 121 |
| **sample()** | base | sample takes a sample of the specified size from the elements of x using either with or without replacement. | 38, 123, 127, 128, 164 |

| Function | Package | Description | Page |
|----------|---------|-------------|------|
| **sapply()** | base | Returns a list of the same length as X, each element of which is the result of applying FUN to the corresponding element of X. | 63, 65, 70, 148 |
| **scan()** | base | Read data into a vector or list from the console or file. | 52 |
| **sd()** | stats | This function computes the standard deviation of the values in x. If na. rm is TRUE then missing values are removed before computation proceeds. | 90, 112 |

| Function | Package | Description | Page |
|---|---|---|---|
| **search()** | base | Gives a list of attached *packages*, and R objects, usually data. frames. | 7 |
| **seconds()** | chron | Return Seconds from a Times Object. | 47, 49 |
| **seq()** | base | Generate regular sequences. seq is a standard generic with a default method. | 20, 45, 46, 64, 92, 93, 94, 95, 106 |
| **seq.int()** | base | Generate regular sequences. seq.int is a primitive which can be much faster but has a few restrictions. | 20, 38 |
| **seq_along()** | base | Generate regular sequences. seq_ along and seq_len are very fast primitives for two common cases. | 20 |
| **seq_len()** | base | Generate regular sequences. seq_ along and seq_len are very fast primitives for two common cases. | 20 |

| Function | Package | Description | Page |
|---|---|---|---|
| **set.seed()** | base | This function is the recommended way to specify seeds for Random Number Generation. | 121, 125, 127, 128, 136, 147, 156, 158, 165 |
| **setRepositories()** | utils | Interact with the user to choose the package repositories to be used. | 8 |
| **setwd()** | base | Used to set the working directory to dir. | 41, 42, 49, 172, 182 |
| **simplify()** | igraph | Creates simple graphs. Simple graphs are graphs which do not contain loop and multiple edges. | 190, 194 |

| Function | Package | Description | Page |
|---|---|---|---|
| **SMOTE()** | DMwR | This function handles unbalanced classification problems using the SMOTE method. Namely, it can generate a new "SMOTEd" data set that addresses the class unbalance problem. | 160, 164 |
| **solve()** | base | This generic function solves the equation a %*% x = b for x, where b can be either a vector or a matrix. | 26 |

| Function | Package | Description | Page |
|----------|---------|-------------|------|
| **sort**() | base | Sort (or order) a vector or factor (partially) into ascending or descending order. For ordering along more than one variable, e.g., for sorting data frames, see order. | 59, 60, 70, 132, 134, 156, 159 |
| **sqldf**() | sqldf | SQL select on data frames | 61 |
| **sqlQuery**() | RODBC | Submit an SQL query to an ODBC database, and retrieve the results. | 55 |
| **sqrt**() | base | This function computes the square root of x. | 6, 143 |
| **stamp**() | lubridate | Stamps are just like format, but based on human-frendly templates like "Recorded at 10 am, September 2002" or "Meeting, Sunday May 1, 2000, at 10:20 pm". | 46 |
| **stemCompletion**() | tm | Heuristically complete stemmed words. | 153 |
| **stopwords**() | tm | Return various kinds of stopwords with support for different languages. | 153 |
| **str**() | utils | Compactly display the internal structure of an R object, a diagnostic function and an alternative to summary (and to some extent, dput). | 51, 54, 144 |

| Function | Package | Description | Page |
|---|---|---|---|
| str_count() | stringr | Count the number of matches in a string. | 57, 70 |
| str_detect() | stringr | Detect the presence or absence of a pattern in a string. | 56, 57 |
| str_replace() | stringr | Replace matched patterns in a string. | 57, 70 |
| str_replace_all() | stringr | Replace matched patterns in a string. | 57 |

| Function | Package | Description | Page |
|---|---|---|---|
| str_split() | stringr | Split up a string into pieces. | 57, 70 |
| str_split_fixed() | stringr | Split up a string into pieces. | 57 |
| strftime() | base | Functions to convert between character representations and objects of classes "POSIXlt" and "POSIXct" representing calendar dates and times. | 44 |
| stripWhitespace() | tm | Strip extra whitespace from a text document. Multiple whitespace characters are collapsed to a single blank. | 153 |

| Function | Package | Description | Page |
|----------|---------|-------------|------|
| **strptime**() | base | Functions to convert between character representations and objects of classes "POSIXlt" and "POSIXct" representing calendar dates and times. | 43, 44, 172, |
| **strsplit**() | base | Split the elements of a character vector x into substrings according to the matches to substringsplit within them. | 41 |
| **structure**() | base | This function returns the given object with further attributes set. | |
| **strwrap**() | base | Each character string in the input is first split into paragraphs. | 153 |
| **sub**() | base | Search for matches to argument pattern within each element of a character vector. | 20,70 |
| **subset**() | base | Return subsets of vectors, matrices or data frames which meet conditions. | 34, 155, 170, 172 |
| **substr**() | base | Extract or replace substrings in a character vector. | 20, 41 |

| Function | Package | Description | Page |
|----------|---------|-------------|------|
| substring() | base | Extract or replace substrings in a character vector. | 41 |
| sum() | base | sum returns the sum of all the values present in its arguments. | 3, 4, 5 |

| Function | Package | Description | Page |
|----------|---------|-------------|------|
| summary() | base | summary is a generic function used to produce result summaries of the results of various model fitting functions. The function invokes particular methods which depend on the class of the first argument. | 45, 46, 67, 91, 99, 103, 104, 1112, 131, 136, 148, |
| switch() | base | switch evaluates EXPR and accordingly chooses one of the further arguments. | 13, 14, 16 |
| Sys.getenv ("HOME") | base | Sys.getenv obtains the values of the environment variables. | 7 |
| Sys.getlocale() | base | Get details of or set aspects of the locale for the R process. | 44 |
| Sys.time() | base | Sys.time and Sys.Date returns the system's idea of the current date with and without time. | 42, 43, 44, 48, 49 |
| Sys.timezone() | base | Information about time zones in R. Sys.timezone returns the name of the current time zone. | 44, 49 |

| Function | Package | Description | Page |
|---|---|---|---|
| **system.file()** | base | Finds the full file names of files in packages etc. | 51, 53, 55, 70, |
| **t()** | base | Given a matrix or data. frame x, t returns the transpose of x. | 25, 34, |
| **table()** | base | table uses the cross-classifying factors to build a contingency table of the counts at each combination of factor levels. | 38, 114, 127, 151, 168, 171 |
| **tabulate()** | base | tabulate takes the integer-valued vector bin and counts the number of times each integer occurs in it. | 90, 112 |
| **tan()** | base | These functions give the obvious trigonometric functions. They respectively compute the cosine, sine, tangent, arc-cosine, arc-sine, arc-tangent, and the two-argument arc-tangent. | 6 |
| **tapply()** | base | Apply a function to each cell of a ragged array, that is to each (non-empty) group of values given by a unique combination of the levels of certain factors. | 63, 66, 67, 68, 69 |

| Function | Package | Description | Page |
|---|---|---|---|
| **TermDocument Matrix()** | tm | Constructs or coerces to a term-document matrix or a document-term matrix. | 154 |
| **text()** | graphics | This function Adds Text to a Plot | 125, 127, 165, 168 |
| **theme()** | ggplot2 | Modify components of a theme | 77, 173, 174 |
| **times()** | chron | Generate Times Components from Input | 171 |
| **tm_map()** | tm | Interface to apply transformation functions to corpora. | 153 |
| **today()** | lubridate | Returns the current date | 46, 47, 48 |
| **toJSON()** | jsonlite | Converts R objects to JSON | 53, 70 |
| **tolower()** | base | Translate characters in character vectors, in particular from upper to lower case or vice versa. | 41 |
| **toString()** | base | This is a helper function for format to produce a single character string describing an R object. | 39 |
| **toupper()** | base | Translate characters in character vectors, in particular from upper to lower case or vice versa. | 41 |

| Function | Package | Description | Page |
|---|---|---|---|
| **union**() | base | Performs the set operation union on the two member vectors | 138 |
| **unique**() | base | unique returns a vector, data frame or array like x but with duplicate elements/rows removed. | 37, 90, 112, 171, 173 |
| **unlist**() | base | Given a list structure x, unlist simplifies it to produce a vector which contains all the atomic components which occur in x. | 30, 153 |
| **update**() | stats | update will update and (by default) re-fit a model. It does this by extracting the call stored in the object, updating the call and (by default) evaluating that call. | 77 |

| Function | Package | Description | Page |
|---|---|---|---|
| **update. packages**() | utils | old.packages indicates packages which have a (suitable) later version on the repositories whereas update. Packages offers to download and install such packages. | 8 |

| Function | Package | Description | Page |
|----------|---------|-------------|------|
| **vapply()** | base | Returns a list of the same length as X, each element of which is the result of applying FUN to the corresponding element of X. | 63, 65, 70 |
| **var()** | stats | Computes the variance of x. | 90, 112 |
| **varImpPlot()** | random Forest | Dotchart of variable importance as measured by a Random Forest. | 129, 149, 151, 166 |
| **vector()** | base | vector produces a vector of the given length and mode. | 20, 49 |
| **VectorSource()** | tm | This function creates a vector source. | 153 |
| **View()** | utils | Invoke a spreadsheet-style data viewer on a matrix-like R object. | 7, 8 |
| **weekdays()** | base | This function Extracts the weekday. | 171, 172 |
| **weeks()** | lubridate | Quickly create Period object (week) for easy date-time manipulation | 47, 49 |
| **which()** | base | Give the TRUE indices of a logical object, allowing for array indices. | 22, 49, 137, 148, 170 |
| **which.max()** | base | Determines the location, i.e., index of the (first) minimum or maximum of a numeric (or logical) vector. | 22, 90, 112, 166 |

| Function | Package | Description | Page |
|----------|---------|-------------|------|
| **which.min()** | base | Determines the location, i.e., index of the (first) minimum or maximum of a numeric (or logical) vector. | 22 |
| **with()** | base | Evaluate an R expression in an environment constructed from data, possibly modifying (a copy of) the original data. | 58 |
| **with_tz()** | lubridate | This function returns a date-time as it would appear in a different time zone. | 48 |

| Function | Package | Description | Page |
|----------|---------|-------------|------|
| **within()** | base | Evaluate an R expression in an environment constructed from data, possibly modifying (a copy of) the original data. | 58 |
| **wordcloud()** | wordcloud | Plot a word cloud | 156 |
| **write.csv()** | utils | Similar to write.table() | 52, 70 |
| **write.foreign()** | foreign | This function exports simple data frames to other statistical packages by writing the data as free-format text and writing a separate file of instructions for the other package to read the data. | 54 |

| Function | Package | Description | Page |
|----------|---------|-------------|------|
| **write.table()** | utils | write.table prints its required argument x (after converting it to a data frame if it is not one nor a matrix) to a file or connection. | 52, 70 |
| **write.xlsx2()** | xlsx | Write a data.frame to an Excel workbook. | 53, 70 |
| **writeLines()** | base | Write text lines to a connection. | 52, 53, 70, 153 |
| **writeMat()** | R.matlab | Writes matlab files | 54, 70 |
| **xmlParse()** | XML | This function Parses a XML file or XML content, and generates an R structure representing the XML tree. | 53, 70 |
| **xmlTreeParse()** | XML | Parses an XML file or string containing XML content, and generates an R structure representing the XML tree. | 53 |
| **xyplot()** | lattice | This function produces bivariate scatterplots or time-series plots | 75, 76, 79, 87 |
| **yaml.load()** | yaml | Parse a YAML string and return R objects. | 53 |
| **yaml.load_ file()** | yaml | This function calls yaml. load with the contents of the specified file or connection. | 53 |

| Function | Package | Description | Page |
|---|---|---|---|
| **years()** | lubridate | Quickly create Period object (year) for easy date-time manipulation | 47, 49, |
| **ymd_hm()** | lubridate | Parse dates that have hours or minutes elements. | 46 |
| **ymd_hms()** | lubridate | Parse dates that have hours, minutes, or seconds elements. | 46 |

# REFERENCES

## BOOKS

1. Paul Teetor, "R Cookbook", O'Reilly Media, 2011.
2. Ohri, "R for Business Analytics", Springer, 2012.
3. Yanchang Zhao, "R and Data Mining: Examples and Case Studies", ELSEVIER, 2012.
4. Richard Cotton, "Learning R - A Step-by-Step Function Guide to Data Analysis", O'Reilly Media, 2013.
5. Roger D. Peng, "R Programming for Data Science", Lean Publishing, 2014.
6. Yanchang Zhao, Yonghua Cen, "Data Mining Applications with R", ELSEVIER, 2014.

## WEBSITES

1. https://www.tutorialspoint.com/r/
2. http://www.r-tutor.com/
3. https://cran.r-project.org/manuals.html
4. https://www.r-bloggers.com/
5. http://www.rdatamining.com/examples/

# INDEX

Made in the USA
Monee, IL
07 July 2026

56550053R00222